About Island Press

Since 1984, the nonprofit organization Island Press has been stimulating, shaping, and communicating ideas that are essential for solving environmental problems worldwide. With more than 1,000 titles in print and some 30 new releases each year, we are the nation's leading publisher on environmental issues. We identify innovative thinkers and emerging trends in the environmental field. We work with world-renowned experts and authors to develop cross-disciplinary solutions to environmental challenges.

Island Press designs and executes educational campaigns, in conjunction with our authors, to communicate their critical messages in print, in person, and online using the latest technologies, innovative programs, and the media. Our goal is to reach targeted audiences—scientists, policy makers, environmental advocates, urban planners, the media, and concerned citizens—with information that can be used to create the framework for long-term ecological health and human well-being.

Island Press gratefully acknowledges major support from The Bobolink Foundation, Caldera Foundation, The Curtis and Edith Munson Foundation, The Forrest C. and Frances H. Lattner Foundation, The JPB Foundation, The Kresge Foundation, The Summit Charitable Foundation, Inc., and many other generous organizations and individuals.

Generous support for this publication was provided by Russell and Carol Faucett.

The opinions expressed in this book are those of the author(s) and do not necessarily reflect the views of our supporters.

The Cougar Conundrum

The Cougar Conundrum

SHARING THE WORLD WITH
A SUCCESSFUL PREDATOR

Mark Elbroch

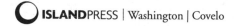

ISLANDPRESS | Washington | Covelo

Note: All images not otherwise credited are by the author.

Library of Congress Control Number: 2020931783

All Island Press books are printed on environmentally responsible materials.

Manufactured in the United States of America
10 9 8 7 6 5 4 3 2 1

Keywords: Island Press, cougar, conundrum, wildlife management, conservation, puma, mountain lion, panther, coexistence, big cats, advocacy, keystone species, ecosystem services, tolerance, predator-prey, carnivores, ecology, predation, depredation, apex predator, biodiversity, ecosystem health, resilience, hunting, harvest, human-wildlife conflict, lethal control, democracy, Pittman-Roberts Act, Teaming with Wildlife, Teddy Roosevelt, Maurice Hornocker, misinformation, social media, attacks, human fatalities, livestock, Torres del Paine, human safety, ecosystem engineer, houndsman, hound hunting, inclusivity, Panthera

For Enzo and Avery, and for
mountain lions everywhere

And I think in this empty world there was
room for me and a mountain lion.
— D. H. Lawrence, "Mountain Lion"

Contents

Preface

Puma, cougar, panther, mountain lion, catamount, mountain screamer, painter, red tiger, león, and leopardo are just several of the more than eighty common names that describe a graceful, stealthy predator that inhabits the largest range of any terrestrial mammal in the Western Hemisphere. Common names remain regional. In Florida, these animals are called panthers, and in the Pacific Northwest, cougars. In much of the remaining West, they are most often called mountain lions, or just "lions" for short. Taxonomists call this species Puma concolor. *The genus name* Puma *means "powerful animal" in an old Incan dialect. The species name* concolor *means "single color" and is meant to describe the uniform pelage of adult animals.*

In late 2017, Steve Ecklund, host of a Canadian hunting television show, *The Edge*, posted public pictures on Facebook of himself and the dead mountain lion he had killed. He is an enthusiastic hunter and sportsman, he's an ambassador for Cabela's, the hunting equipment retail giant, and he'd posted similar images of other animals he'd hunted many times before. But unexpectedly, this one caused a surge of public backlash that even included Laureen Harper, the wife of the former prime minister. She took the time to post this on Twitter: "What a creep. . . Must be compensating for something, small penis probably." It's a low blow to attack a man's genitalia.

I barely registered the spike in social media surrounding Ecklund and his trophy mountain lion, excepting that friends kept sending me links to the various articles about the incident. The year 2017 had already been full of sensational news about mountain lions attacking pets and livestock, as well as warnings about protecting children following mountain lion sightings in various Western neighborhoods. One lion was killed in urban downtown Salt Lake City—that incident raised some eyebrows—and there had been real excitement around a YouTube video of an encounter with a mountain lion on the High Sierra Trail in California.

A pair of hikers glimpsed a mountain lion on the trail before them. That wasn't unusual, but then they pursued it when it trotted away from them. Around the next corner, they found themselves much too close to the animal, which had lain on the slope above the trail. The encounter was peaceful and the cat did not show any signs of aggression that I could see in the videos I perused. Nevertheless, there were the inevitable debates about whether the hikers had acted appropriately or been foolish and were lucky to have survived.

There is a near-constant back-and-forth on various social media outlets between groups of individuals with different opinions about mountain lions, and nowadays I only half listen—it's like the static you hear on a handheld receiver, which every biologist learns is a distraction from the real signal indicating the direction of the animal you are looking for.

At the same time that Ecklund found himself the center of a cyberstorm, I was trying very hard to block out any lurid news about mountain lions so I could write a book on the little-known social behaviors of the species. Simultaneously, however, the Humane Society of the United States was initiating a signature campaign to end bobcat and mountain lion hunting in Arizona, and Safari Club International was petitioning the United States Court of Appeals for the Ninth District to reverse the ban on mountain lion hunting in California. These developments were more difficult to ignore. They represented bolder, more-strategic efforts to impact mountain lion management that were being orchestrated by lobby groups representing different private-citizen stakeholders. Both efforts failed, but I followed them closely.

In early 2018 there was a further distraction when the United States Fish and Wildlife Service (USFW) announced the extinction of the eastern cougar, a subspecies of mountain lion. I found myself sucked into debates and confusion over the ruling and its implications for mountain lions across North America. The truth is that recent genetic research has shown that there never was a distinct "eastern cougar" in the first place, and that the media was emphasizing all the wrong aspects of the ruling. Then in May 2018, a mountain lion attacked two mountain bikers in Washington State, killing one. A fellow researcher dispatched the animal, which was a young male grossly underweight but otherwise healthy. It was the first mountain lion fatality in Washington State in some ninety years, but the media spread the story and its associated drama from coast to coast. Then it was back to pet and livestock attacks and more fear-mongering public warnings about mountain lions that had been spotted in Western neighborhoods, some near schools. By the time authorities in Oregon had announced the heartbreaking news that a mountain lion had killed a woman in Mount Hood National Forest in September, I'd given up writing a book about mountain lion social behaviors. She was the first person ever killed by a mountain lion in Oregon, but given that she was the second person killed by a mountain lion in the United States in 2018, the incident resulted in renewed debate over mountain lion hunting. Every mountain lion story in the news eventually comes down to hunting.

Instead of writing on social behaviors, I kept finding myself penning notes to address the mix of facts and misinformation about mountain lions and their management found on social media, facts and non-facts used like bludgeons by both wildlife and hunting advocates to quell their opposition. Or I'd find myself trying to explain the political arena in which mountain lion management operates, which serves as a starting place for discussing how hunting impacts the social organization of the species.

People are angry. We—and I include myself—are frustrated with the state of things. We want answers, and more importantly we want to be heard, to feel that those in power listen to us as people invested in wildlife and its management. How many mountain lions are there? Are they really killing all the elk and deer? Are they really a risk to our pets, our livestock, or our children? Do we really need mountain lions in natural systems in the first place?

I decided that I can always write a book about mountain lion social behaviors. What was needed now was a guide to sifting facts from sensationalism. The frenzy of media activity surrounding mountain lions in 2017 and 2018 was not new, just more of what has become the new normal. Time and again, people have risen up to condemn mountain lions and call for increased control of their populations, if not their complete eradication. In response, the opposition has risen up to defend the cats, blaming humans and our encroachment into mountain lion habitat for every tragic incident. Certainly the details of each mountain lion story in the news were different, but the buzz, the anger, the frustration, and the hateful rhetoric wielded by opposing sides have been redundant again and again. They were not mere reactions to any single happenstance, but symptoms of a much larger problem we currently face in North America: the *cougar conundrum*.

In 2016, P-45 was a young male mountain lion living in and around Malibu in the Santa Monica Mountains northwest of Los Angeles, California. He had a name and a number because he wore a GPS collar and was being followed by Seth Riley and Jeff Sikich, scientists for the National Park Service. They researched the local mountain lion population and its chances for survival, given its isolation from other mountain lion populations to the north and southeast.

They predict the Santa Monica mountain lion population will go extinct unless we can build wildlife bridges over Highway 101 to facilitate connectivity with other mountain lions to the north. The mountain range is hemmed in by the Pacific Ocean on one side and southern California suburban sprawl on the other, limiting opportunities for any mountain lion to come or go from the tiny 275-square-mile mountain range. Worse, the area has been made an island by the ten-lane highway that runs its length to the north.

Today, researchers and wildlife advocates work to save a mountain lion population that, just sixty years previously, was likely extirpated following state-paid bounties for dead mountain lions. Between 1907 and 1963, California paid out 12,461 bounties for mountain lion carcasses.[1] During the first half of the twentieth century, the state employed up to five people dedicated solely to the hunting and killing of mountain lions. Jay Bruce was among them, credited with killing more than 700 California mountain lions over twenty-eight years of service.[2]

Even though LA's entertainment industry dominates local culture, southern California is a difficult place for most mountain lions to gain media attention. That is because, in LA, P-22 reigns as king. P-22

P-22, an adult male mountain lion with a Facebook page, walks beneath the iconic Hollywood sign in Griffith Park—an emblematic image of urban mountain lions living on the edge in northern Los Angeles, California. (Photograph courtesy of Steve Winter / *National Geographic*)

somehow miraculously survived crossing eight lanes of commuter traffic on the 405 and eight to ten lanes of traffic on the 101, both mammoth highways with dense traffic at every time of day and night. He was discovered inhabiting the tiny urban wilderness of Griffith Park in 2012, when a local biologist, Miguel Ordeñana, caught an image of him on a motion-triggered camera. "It was like finding an image of Bigfoot or the *chupacabra*," Ordeñana later explained.[3]

P-22 was immortalized in 2013 when *National Geographic* photographer Steve Winter captured an image of him walking beneath the iconic Hollywood sign. Three years later, however, P-45 grabbed the limelight for a while, but for all the wrong reasons: in a single night he killed ten alpacas on a ranch beside the Mulholland Highway in Malibu.

Having lost alpacas to lions in the past, the ranch was not without its defenses. Victoria Vaughn-Perling had strung up solar-powered lights and added flagging to the top of the fence enclosing the animals, in order to give the illusion that it was taller. She even rigged up a radio to blast talk radio, just to keep her livestock safe. But on Thanksgiving in 2016, heavy rain destroyed Vaughn-Perling's defenses, shorting the radio and stopping the lights from recharging. In the silence and darkness that resulted, P-45 slipped in and for some reason killed more than he could eat—unusual behavior for a mountain lion. Vaughn-Perling, discovering the carnage the next morning, contacted the California Department of Fish and Wildlife and requested a depredation permit to pursue and legally kill the lion. She'd had enough; it was time to remove the perpetrator.

The result was an eruption of public outcry as had never been experienced for mountain lions anywhere. On the one side were angry livestock owners, led by a colorful character named Wendell Phillips. who was neighbor to Vaughn-Perling and had earlier wounded P-45 after the lion killed an alpaca on his property.[4] Livestock owners presented their case as follows: 1) there were too many mountain lions prowling the area; 2) P-45 was dangerous to people as well as animals; 3) P-45 was abnormal and seemed to enjoy killing; and 4) P-45 should be removed permanently. Phillips called P-45 "the John Wayne Gacy of mountain lions," referring to a notorious 1970s serial killer.

On the other side, animal rights and mountain lion advocates argued that 1) livestock owners were to blame for livestock conflict; 2) proper enclosures for livestock would solve everything; and 3) P-45 was too valuable to the local mountain lion population to lose. Mountain lions were, in fact, a species of special concern and had been awarded full protection in the Golden State, and their population in the Santa Monica Mountains was particularly vulnerable. Local advocates knew a great deal about the state of local mountain lions because there was active research being conducted on the population. Seth Riley and Jeff Sikich, the biologists who tracked P-45, supported their arguments, explaining that P-45 was one of just two resident male mountain lions in the entire mountain range at the time, and one that had somehow traveled across the 101 from the north to infuse the local mountain lion population with new genetic material. P-45 was therefore a critical animal to conserve in the Santa Monica Mountains.

Vaughn-Perling and Phillips subsequently received death threats from militant wildlife advocates for requesting a permit to kill P-45.[5] In response, livestock owners retorted that they would not just shoot lions but also any people they found skulking on their properties, and discourse on social media deteriorated rapidly. Most mountain lion incidents stop there, with everyone fighting, but the Malibu community rallied and conducted productive public and private meetings about what to do about P-45 and local mountain lions more broadly. The discussions were at times heated and loud, but everyone had the opportunity to speak their mind. Under public pressure, many argue, Vaughn-Perling changed her mind and did not pursue her right to kill P-45 after he killed ten of her alpacas on that terrible night. She told media that her intention had always been that the state would relocate the animal, not kill it.

The results of P-45's carnage, the international outrage that followed, and the public meetings about what to do in response were overall incredibly positive, even if unsatisfactory to all involved. There were a flurry of letters to the editor of the *Los Angeles Times*, some defending the livestock owner's right to kill the animal, and others calling for non-lethal solutions to the conflict. In the end, Vaughn-Perling erected four ten-by-ten-foot chain-link enclosures, and now every night she shepherds her remaining alpacas inside these structures to shield them from P-45 and other lions. The enclosures were designed, supplied, and built on site by the nonprofit Mountain Lion Foundation, and paid for by the National Wildlife Federation, in what many tout as role-model support from mountain lion advocates to put their money where their mouth is and provide feasible solutions to livestock conflict. P-45 still roams and contributes to the local mountain lion population, occasionally killing less-protected livestock. P-45 and other lions will inevitably continue to kill vulnerable livestock in the region unless similar defensive measures are undertaken more broadly. This is part of the reality of living with livestock in mountain lion country.

In 2018, California changed its policy to a "three-strike rule" with regard to issuing depredation permits in the region surrounding the Santa Monica Mountains, in order to further protect the remaining mountain lions there. Owners are now required to try nonlethal methods two separate times to deter repeat livestock killing, and only if that fails will the state wildlife agency issue a permit to kill the offending lion. Mountain lion advocates were thrilled with this outcome, but some livestock owners were outraged. Wendell Phillips said it will encourage vigilante justice. "The reality is nobody will bother to apply for permits anymore," he predicted. "Shoot, shovel, and shut up—that's what coming."[6]

Nevertheless, I would argue that, at least on a local scale, the Malibu community wrestled with the cougar conundrum and came up with a solution. Not everyone involved was happy with the end result, but it is the process that's most important, not the outcome. People from opposing sides participated and contributed to a solution that worked for the livestock owner who suffered losses and also for the larger community.

A century ago, we tried very hard to eradicate mountain lions from the United States and Canada, and we failed. We did succeed in most of the East, except for a few stragglers hiding in Florida swamplands. Some of these elusive, intelligent animals, however, retreated to mountain strongholds in the West and thwarted our every effort to eliminate them. They survived for so long that American culture began to change and to articulate an appreciation for predators as part of healthy ecosystems. Aldo Leopold, for example, stopped killing wolves and began promoting them. What followed was the end to the bounty hunting of mountain lions, and a switch to managing mountain lions with fixed hunting seasons, which imposed limits on the number of animals that could be killed each year.

Then something miraculous happened, something no one would have predicted at the time. The mountain lions that survived the onslaught of previous decades emerged from hiding and began to find each other and mate; their populations rebounded with unexpected vitality and quickly spread into areas where they had previously been exterminated. In parts of the West, mountain lions may now be as abundant as they were in historic times before European settlement, and in some areas, they might be *more* abundant. Mountain lions are, without doubt, among the most successful conservation stories of the last century, and yet few people seem to appreciate this fact or even to acknowledge it. We actually helped a species bounce back from near-extinction, and it is now, on its own, gradually recolonizing areas where it was eradicated a century ago.

The heart of the cougar conundrum is this: Can we peacefully coexist with such a successful predator? The answer is yes, of course we can. More to the point, the question is: Will we choose to peacefully coexist with such a successful predator? That is a question I cannot answer, but given the increasing polarization surrounding mountain lions—the us-versus-them so visible in media surrounding mountain lions and reflective of larger patterns in American culture—the more likely it is that we will remain mired in power struggles and minor social-media fights while avoiding the core issues altogether.

Today, there are more mountain lion advocates and advocacy organizations than ever before, while at the same time, we are killing record numbers of mountain lions in the United States each year. We kill scores of mountain lions to aid bighorn sheep recovery across the Southwest, while simultaneously Los Angeles residents are raising $87 million to build a bridge spanning ten lanes of pavement to save P-45 and a handful of mountain lions in the Santa Monica Mountains. State and federal agencies fund the killing of predators that kill livestock, but they don't fund infrastructure to protect livestock. These are ironies reflective of the quandary we face.

This book is a detailed accounting of the current conundrum, or more specifically, a discussion of a collection of conundrums borne of rebounding mountain lion populations. The book's chapters are thematic, each focusing on a single conundrum. Chapter 1, for example, begins with "What is a mountain lion?" In 1906, President Theodore Roosevelt wrote, "No American beast has been the subject of so much loose writing or of such wild fables as the cougar."[7] More than a century later, the same might still be said. Theodore Roosevelt became uniquely involved in the national debate over the ecological accuracy of the published literature, especially the materials being provided to educate students of the era. He had grave concerns about authors who broadly became known as the Nature Fakers—those who fabricated myths about the workings of the natural world. Mountain lion mythology borne of the nineteenth and early twentieth centuries remains the core of the cougar conundrum.

This book, however, is more than straight documentation. Somewhat reluctantly, I step beyond the comfort zone of a working mountain lion biologist and provide bold speculation and suggestions about how we might escape becoming mired in the conundrum. Chapters 2 and 3 address human and livestock safety issues. Chapter 4 discusses the influence of mountain lions on the ungulate populations so valued by hunters, as well as the state agency programs that serve them. Chapter 5 assesses how mountain lions are faring in a modern world, and chapter 6 explores eastern mountain lions, from sightings to migrants to Florida panthers. Chapter 7 is a discussion of the tactics we might employ to build social capital for mountain lions. Everything culminates in chapter 8, in which we explore who is making and who *should* be making decisions about mountain lion management. More to the point, it is a story of the money and power dynamics that drive wildlife management. Chapter 9 is a discussion of what might come next, depending upon the choices we make now.

The components of forward movement seem quite plain, even if complicated by power dynamics, outdated ideals, and entrenched bureaucracy. First, we must weed out myths and share an understanding of the real mountain lion. Second, we must answer people's questions and concerns about mountain lions. Third, we must come to some consensus around the legal hunting of mountain lions, as well as what to do when a mountain lion kills pets or livestock, or in the rare cases when one attacks a person. Fourth, we must exercise inclusive and transparent mountain lion management. Parallel with these milestones, we must also foster greater tolerance for mountain lions (and large predators more broadly) in our culture and around the world. We must choose how we *live with mountain lions*—essentially, how we might share natural resources like habitat, deer, and elk with them. Myself, I hope we choose peaceful coexistence over any other alternative.

People with extreme views are currently driving the discourse about mountain lions, and we need to be reminded that these views are not representative of most people. Like cornered mountain lions, people sometimes bare their claws when they find themselves on the defensive. Devolving into a social-media troll and engaging in mudslinging on Twitter does not help; in fact, that perpetuates the problem. We are a proudly democratic society. All people are entitled to opinions, and each perspective is valid. Livestock losses, for example, are tragic and hard on real people.

Take a deep breath and let down your defenses. Enjoy the stories, be open to learning something new about mountain lions. My goal was to write a book about these animals that no one would throw down in disgust, regardless of their values and beliefs. Instead, I've tried to be brutally honest, to present the facts as I see them, and to tell stories that educate and entertain. We must bridge the divides that separate the many factions invested in mountain lions, and work together to ensure a future that includes healthy populations of mountain lions and people. Malibu did it, and so can we on a national and international scale.

Acknowledgments

Thank you to Erin Johnson and everyone at Island Press for the opportunity, encouragement, and coaching that resulted in this book. Erin in particular made every effort to provide me with guidance in how I should approach writing. Once I understood the structure, the words just spilled out of me. Thank you to Anne Hawkins for continued support and faith and determination, and for seeing this project through until it became a reality. Thank you to my parents and Shawna Bebo, without whose support, at home and from afar, I'd never have been able to juggle the responsibilities of family, writing, and work. Thank you to Fred Launay and Howard Quigley, my supervisors at Panthera, for granting me the opportunity to write this book while also managing my other commitments.

A thousand thank-yous to the mountain lions that have endured my attention. They have enriched, and continue to enrich, my life beyond description.

My path with regard to mountain lion research and conservation has been long and winding, and I am grateful for so many wonderful people who have shared time with me in the field, provided mentorship in science and fieldwork, and/or provided inspiration, even if from afar. All of these people have in some way shaped me and this book. In alphabetical order, I would like to thank the following individuals (and I apologize to anyone I may have overlooked): Pete Alexander, Max Allen, Casey Anderson, Melanie Anderson, Stephanny Arroyo-Arce, Josh Barry, Rich Beausoleil, Paul Beier, Steve Bobzien, Randy Botta, Walter Boyce, Jorge Cardenas, Sara Cendejas-Zarelli, Justin Clapp, Susan Clark, Paulo Corti, Derek Craighead, Bogdan Cristeau, Justin Dellinger, Roberto Donoso, Holly Ernest, Jen Feltner, Jenny Fitzgerald, Tavis Forrester, Dania Goic, Melissa Grigione, Jeff Hogan, Rafael Hoogesteijn, Maurice Hornocker, Jen Hunter, Ken Jafek, Brian Jansen, Doug Kelt, Marc Kenyon, Brian Kertson, Anna Kusler, John Laundré, Sharon Lawler, Patrick Lendrum, Ken Logan, Carlos López-González, Blake Lowrey, Cameron Macias, Dave Manson, Quinton Martins, Dave Mattson, Casey McFarland, Blue Milsap, Mauricio Montt, Rodrigo Moraga, Chris Morgan, Sharon Negri, Matt Nelson, Jesse Newby, Omar Ohrens, Connor O'Malley, Miguel Ordeñana, Megan Parker, Michelle Peziol, Mike Puzzo, Howard Quigley, Seth Riley, Cristián Rivera, Lisa Robertson, Hugh Robinson, Drew Rush, Toni Ruth, Kim Sager-Fradkin, Ron Sarno, Cristián Saucedo, Mark Schwartz, Ashleigh and Gerry Scully, Don Arcilio Sepúlveda, Harley Shaw, Mike Sheldon, Andy Sih, Jeff Sikich, Boone Smith (who has shared more cat adventures with me than any other, and years of friendship), Sam Smith, Luigi David Solis, Jack and Jodie Stevenson, Andy Stratton, Don Strong, Linda Sweanor, Dan Thompson, Steve Torres, Fernando Tortato, Winston Vickers, Neal Wight, Jeremy Williams, Chris Wilmers, Steve Winter, Heiko Wittmer (my graduate adviser and continued partner in conversation), and Eric York, one of the best field biologists I've ever met.

CHAPTER 1

"The Lord of Stealthy Murder" and Other Misconceptions

A pair of snow mobiles ripped through the mountains northeast of Jackson, Wyoming, at dawn, their whining engines a profanity in the still silence of that wild place. Patrick Lendrum and Michelle Peziol were driving, clad in one-piece suits and helmets that covered and protected every inch of skin from the bitter cold. They were biologists for Panthera's Teton Cougar Project, and the project included some unusual methods. GPS collars worn by mountain lions relayed the locations of the animals in near-real time to the project's computers via satellites orbiting the Earth. Patrick and Michelle spent their days racing to set motion-triggered cameras in areas where mountain lions lingered long enough that their behaviors could be recorded. They did so even knowing they might encounter a mountain lion in the remote terrain.

When Michelle and Patrick arrived on site that morning, they found the snowmobile trail stippled with fresh lion footprints. Michelle set to stomping out M58's footprints in order to hide his presence from humans who might harass him, or worse. Patrick dropped away from the groomed snowmobile trail and waded through deep snow toward

An intimate portrait of an adult female mountain lion in northwest Wyoming.

open water. The Gros Ventre River raged nearby, loud and raucous, rolling the rocks in exposed rapids rushing between patches of ice. Patrick spotted the hind end of an elk protruding from the icy water not twenty feet away from him, the legs oddly vertical as if the animal had frozen in position before it was pulled from the river. He yelled to Michelle that he'd found the kill, but the noise of the river swallowed his words.

Michelle worked until she was satisfied that no trace of lion remained, and then she moved toward the river to join Patrick. She stepped off the trail and—*thunk*—sunk up to her waist in deep, soft snow. She shrieked in merriment, and while still laughing at her own predicament, she looked toward Patrick. He had moved along the exposed shoreline to a large Engelmann spruce about ten feet from the carcass—an ideal place to set a camera, she thought. Her smile faltered and fell, however, when she locked eyes with M58 staring at her from the base of the tree. Adrenaline surged through her body. Time slowed.

The lion stood elevated four feet above the ground on a nest of tangled roots, exposed by high waters in years past. He'd been lying below the overhanging foliage of the spruce tree, which provided a roof while he rested. M58 was a curved 4¾ feet (144 cm) from the tip of his nose down along his spine to the notch where his tail connected to his body. The bony parts of his tail added another 2⅔ feet (80 cm). He was robust, barrel-chested and strong. When the team had caught him four months earlier, he'd weighed 165 pounds, a typical adult male weight in the region.

Patrick stood on river rocks below M58, oblivious to the predator behind him. The mountain lion stood motionless, a silent statue, ears erect and forward, studying Michelle, not Patrick. But he was so close to Patrick that he could have reached out a paw and placed it on his shoulder. Michelle whispered to Patrick, loudly enough to project her voice over the river, "There's a lion!" She gestured over his shoulder, keeping her movements discreet. Patrick smiled, assuming she was joking. She whispered a second time, "There's a lion." Patrick studied Michelle more closely. There was no humor in her, and as realization dawned, the color drained from his face. He couldn't help himself. Slowly, he turned toward the trunk of the tree.

Patrick's motion may have well been a gunshot—such was M58's transition from statue-still to blurred motion. One moment he was standing over Patrick; Michelle blinked; and the next he was liquid grace. He leapt from the bed, hit the river edge, and with two more massive bounds entered and cleared the river, landing on the far side. He disappeared into sparse woods before either of the biologists had time to take a breath. There was no time to react, no time to think. M58 was there and then he was gone—just like that.

Patrick and Michelle exhaled loudly, the adrenaline accompanying the urge to fight or flee beginning to dissipate. They both said something like "Wow" or "Whoa," the sudden rush having distorted some details of the encounter. They speculated that M58 may have woken up when Michelle plunged into the deep snow and released her trademark shriek—such was the noise of the river that he hadn't heard the approach of the snowmobiles, or perhaps he'd just ignored them.

In fact, Michelle and Patrick were rarely confronted with mountain lions while they worked, and their encounter with M58 was a surprise. But if you've read books with titles like *Cougar Attacks: Encounters of the Worst Kind*, *Cat Attacks: True Stories and Hard Lessons from Cougar Country*, or *Stalked by a Mountain Lion*, it hardly seems safe practice to approach areas where there are known to be wild mountain lions. Should Michelle and Patrick have been afraid? It is difficult to know what to believe these days, given that misinformation circulates in printed and Internet media faster than it can be debunked.

The story of the European settlement of North America and the decimation of its wildlife is familiar by now and needs little review. Colonists established themselves quickly along the Eastern Seaboard, clearing forests, killing Native people and native animals, and introducing cattle, which increased conflicts with native predators. The first colonists were farmers, not hunters or soldiers. But they still knew how to handle the few weapons they had, and how to harvest game from the land. New settlers were so quick to wipe out local resources that, by 1639, colonies were establishing moratoriums on deer hunting to help local populations recover.[1]

As the eastern coast of North America was transformed and tamed, pioneers and mountain men began moving west, at first in dribs and drabs but eventually in mass migrations following the Revolutionary War and the War of 1812. As along the East Coast, western lands were cleared, and Native people and predators exterminated. Initially, wolves and bears were targeted over mountain lions. Pennsylvania, for example, began paying wolf bounties in 1683 but only offered mountain lion bounties beginning in 1807.[2] Poisons too, were less effective in killing mountain lions, since lions only rarely fed on the bait placed by hunters. But poisons proved more effective as livestock spread west. Mountain lions kill large prey, hide it, and then return to feed from the carcass repeatedly. Thus, if an early pioneer located a lion's kill, which they were more likely to do if the lion killed livestock, they had only to lace the carcass with strychnine to bring about the lion's demise.[3]

As wolves and bears were eradicated, people identified mountain lions as the principal predatory threat in the American West. This perception was encouraged by local stockmen's and sportsmen's associations, both of which were growing in number and political strength in the 1800s.[4] Cattlemen proved to be a triple threat. First, throughout much of the nineteenth century they purchased prime US lands that could have supported wildlife in shady privatization of public lands. Second, they denuded vast tracts of the remaining public lands with unregulated cattle. And third, they convinced the federal government to institute widespread predator control. By 1930, the US Biological Survey (then the Bureau of Biological Survey) employed more than 200 professional cat hunters to aid ranchers suffering losses to mountain lions.[5] State agencies employed lion hunters as well, and a greater number of individuals worked like mercenaries for government bounties, carving out heroic reputations for themselves while killing predators across the West.

These bounty hunters and their government-paid counterparts were the early hound-hunters. They earned their keep by hunting cats in rugged terrain, each killing hundreds of mountain lions, and in a few cases many more. A gentle, soft-spoken British Columbia man named

John Cecil Smith, later dubbed "Cougar Smith," was thought to have killed more than 1,000 lions over sixty years. "Cougar Annie" and Bergie Solberg, who was featured in the film *The Cougar Lady of the Sunshine Coast*, were both women of British Columbia who traded pelts for cash.[6] In November 1918, Jay Bruce became California's first "official lion hunter," a position he held for twenty-eight years. He was estimated to have killed 700 lions over thirty years—many with a revolver bearing the inscription "To J. C. Bruce, State Lion Hunter, California 1919." As was typical at the time, Bruce's work was admired as a significant contribution to society: "His success as cougar killer increased the deer population in his state and saved the livestock of many farmers."[7]

Ben Lilly was a giant even among rugged mountain men. During his lifetime Lilly likely killed more than 1,000 mountain lions, many of which were dispatched with one of his double-edged "Lilly" knives that he forged himself and tempered in "panther oil."[8] He was buried near Silver City, New Mexico, under the epitaph, "Lover of the Great Outdoors."[9] Bruce, Smith, Lilly and a host of other hunters were both proficient and efficient, scouring landscapes for the last big predators.

Simultaneously with systematic predator killing, sportsmen's associations established by wealthy, politically connected men successfully lobbied for increased regulation of hunting and trapping to ensure the sustainable management of the game species they valued. Deer and elk populations began to grow, setting the stage for the recovery of the predators that ate them. Wolves and grizzly bears had been successfully extirpated across the Lower 48 except in a few wild areas, whereas mountain lions had proven more adept at hiding and biding their time. Mountain lions clung to life in the wildest, most rugged parts of the West, areas where only the most tenacious hunters pursued them. And so for mountain lions, the recovering deer and elk populations spelled opportunity.

Even better for mountain lions, the American people began to evolve as well. In contrast to the Jay Bruces and Ben Lillys, charismatic, colorful characters such as John Muir, Aldo Leopold, and Ansel Adams emerged to advocate for the protection of wild places and wild creatures in the first half of the twentieth century. In 1962, Rachel Carson wrote in *Silent Spring* that "The question is whether any civilization can wage relentless war on life without destroying itself, and without losing the right to be called civilized."[10]

State and federal agencies responsible for managing wildlife began experiencing an uptick in external pressure from a more diverse American public. Historically, agencies had dictated wildlife management with little input from the American public, other than from lobbyists paid for by livestock and sportsmen's associations. This change was particularly important for mountain lions, as it was external pressure from the American public rather than agency biologists that was primarily responsible for transitioning mountain lion management from one in which 80 percent of states still paid bounties for beautiful big cats in 1950, to one in which nearly every Western state had replaced bounties with managed hunting limits in 1965.[11]

Margaret Owings, for example, led the charge to end mountain lion bounty hunting in California. Alice Schultz, Leona Miller, and other "erudite and tenacious" women did so in Arizona, advocating for the abolishment of mountain lion bounties at every meeting of the Arizona Game Protective Association, the Arizona Cattle Growers Association, and the Sierra Club. In Arizona in 1971, these "meddling little old ladies in tennis shoes,"[12] as their adversaries referred to them, succeeded in abolishing bounties and establishing regulated hunting seasons for mountain lions.

Without doubt, the widespread reclassification of mountain lions across the West from vermin to "game" species is what saved the species in the Unites States. With game species status, people could no longer kill mountain lions at any time in any fashion, nor could they kill as many mountain lions as they liked. Hunters were required to buy licenses, and they were strictly limited to hunting during seasons monitored by state officials.

Nevertheless, game species status also created new challenges. State wildlife agencies were suddenly charged with maintaining sustainable mountain lion populations for their ecological value and for recreational hunting—in perpetuity. Dr. Wain Evans, assistant director of the New Mexico Fish and Game Department at the time, lamented the fact that game agencies were unprepared to manage mountain lions. "When they did this," he wrote, referring to the change from vermin to game species, "they gave us an animal about which we knew nothing, and at the time not very many other people knew much about the lion either."[13] As of 1970, only Idaho had initiated a mountain lion study to learn more about them.

In the past sixty years, there have likely been more than 100 mountain lion studies amassing tremendous amounts of data on mountain lion biology—what they eat, how far they wander, how often they breed. But we still don't understand these animals. They are cryptic and elusive and rarely observed. For these reasons, houndsmen and -women who hunted mountain lions both provided the earliest accounting of their behavior and defined our cultural understanding of the animal. Their dramatic renditions of cornered mountain lions continue to fuel sensationalized media accounts of mountain lion encounters, because actual facts are lacking.

So, let us begin by re-introducing the mountain lion and reviewing what we know about them. The picture that is emerging is of a successful, adaptable, and resilient animal. Mountain lions are intelligent and curious, but cautious. They are generally peaceful beasts, even when provoked, and evidence suggests that they are willing to live alongside human beings if given the chance.

A young male mountain lion; this lateral view emphasizes the powerful back end and the long, thick tail.

Mountain lions are long and lithe, narrower in the chest than other similar-sized cats from other parts of the world. They have great, thick, sweeping tails with black tips. Their fur coats are soft and thin, and their tawny color is mostly uniform from nose to tail. Most lions wear a darker stripe of fur following their spine, and all are paler in color beneath, approaching white. Adult mountain lion faces are more variable, like painted masks, but all have white "chops" between their nose and mouth, and black-and-white "eye spots," called *ocelli*, on the backs of their ears. Male mountain lions are stockier, and females have slender necks and small heads in contrast to the incredible strength of their hind ends. Males in the United States tend to weight 125–180 pounds, depending upon locale, and sometimes slightly more. Females tend to range between 75 and 100 pounds, occasionally more.

The mountain lion's scientific name is *Puma concolor*. *Concolor* means "single color" and describes the uniform pelage of adult animals. Individual mountain lions are indeed generally one color; however, not all mountain lions are the same color. In Wyoming and Idaho, mountain lions range from ochre to tawny to sepia, and in California the cats are pale yellow to golden brown. In the Pacific northwest, lions are darker shades of brown. In every place, mountain lion colors mimic local earth, vegetation, or dry grass. In Patagonia, they can be gray like a granite boulder, rather than any color of grass or ground. The southernmost mountain lions look like ghosts of their northern counterparts.

On very rare occasions, a mountain lion is pure white—not an albino with red eyes and pink skin, but a "leucistic," white-clad cat devoid of color though with otherwise normal features. But they are never black. In the thousands upon thousands of mountain lions spotted, captured, killed, and studied from Alaska south to the southern tip of South America over the last 300 years, not one has ever been black. The "black panther" is a black leopard or jaguar. The black mountain lion is an animal of myth and imagination, however pervasive.

People generally know what a mountain lion looks like, even if some still misidentify them in the media and on the Internet. The question with which we still grapple is: What makes them tick? Even today, there is a dearth of information on the subject. Should you turn to old accounts, you might suspect there were two different mountain lion species roaming North America. The first creature was a "ferocious and bloodthirsty," "dangerous and savage,"[14] "murderous,"[15] "slaughter hungering"[16] "master killer,"[17] with "evil yellow eyes"[18] that "do not appeal to any feelings of compassion."[19] Worse, it had "teeth sharp as spears along his gory, life destroying snout"[20] and "a heart both craven and cruel."[21] Among its many appellations were mountain devil, devil cat, sneak cat, and mountain demon.[22] The second animal, by contrast, was "a cur,"[23] "mean and thievish . . . a beast of low cunning,"[24] and one "uniformly cowardly,"[25] that was "less courageous than the true [African] lion," with vocalizations that are "a little contemptible."[26] A correspondent of Theodore Roosevelt wrote in a letter that "They are very similar in size and are equally as well armed with teeth and claws, but they have not for a moment the *courage* or *audacity* of the leopard or panther [jaguar]."[27]

Both the dangerous mountain lion and the cowardly one were, for the most part, fabricated by those who hunted them with hounds, cornering cats so that they hissed and snarled and fought as their last line of defense. The dangerous lion followed popular literature that sensationalized the battle of man versus nature and celebrated independent rugged explorers like Daniel Boone. The cowardly lion was born of civilized, wealthy sportsmen who accompanied lion hunters, disdaining an animal that provided little challenge or drama once it was cornered and easy to kill. Some authors of the nineteenth and early twentieth centuries seemed unsure which animal to depict, so they described both. President Theodore Roosevelt, who took great interest in mountain lion hunting, was one such author: "The cougar is a very singular beast, shy and elusive to an extraordinary degree, very cowardly and yet bloodthirsty and ferocious."[28]

The frothing, hissing image of cats cornered by baying hounds has for too long dominated our perceptions of mountain lions.

The lion hunters themselves seemed to think that mountain lions were rarely dangerous, regardless of how they portrayed them to the public in their storytelling. After a decade of partaking in guided mountain lion hunts during which he soaked up the lore of experienced houndsmen, President Teddy Roosevelt wrote, "There is no more need of being frightened when sleeping in, or wandering after nightfall through, a forest infested by cougars than if they were so many tom-cats."[29] He went on to explain, "It is absolutely safe to walk up to within ten yards of a cougar at bay, whether wounded or unwounded, and to shoot it at leisure."[30]

Hunters remained the primary source of mountain lion natural history until the 1960s and '70s, when biologists began to study the species more intensively. Maurice Hornocker is largely credited with being the grandfather of mountain lion science. In 1964, he launched his landmark research of mountain lions in a remote corner of Idaho now designated the Frank Church–River of No Return Wilderness, providing the first comprehensive study of the species ever conducted. Hornocker and his team, including John Seidensticker and houndsman Wilbur Wiles, were the first to estimate the size of a mountain lion's territory and to compare how males and females were distributed on the landscape. They were also the first to conduct an intensive study of mountain lion predation on deer and elk. But Hornocker's research did little to elucidate the personality of mountain lions. It was instead a study of the many parts that make up the behavior of mountain lions.

The first person to attempt a gestalt approach to understanding mountain lions might be naturalist and writer Ronald Douglas Lawrence. Lawrence was skeptical of what could be learned with radio telemetry, and so, flouting traditional scientific approaches, he set forth in the fall of 1972 to engage mountain lions on their own terms. He entered the Selkirk Mountains of British Columbia via canoe on the

Goldstream River, with camping gear, heater, stovepipe, ten books, and 1,100 pounds of supplies.[31]

Lawrence relied upon footprints and other signs to locate his quarry. With the first snows, he began following a particular male mountain lion, ever hopeful of encounters. With persistence, he did indeed meet this mountain lion, catching up to it unexpectedly where it rested one day. He had to master his own fear as the animal scrutinized him and his intentions from a safe distance away. He decided to sit down in the open where his movements were easy to track, allowing the mountain lion to become accustomed to his presence. He deliberately turned his back on the animal, after which the cat moved into a more comfortable position to watch the man that had chosen to approach him.

He used the same tactic during his second encounter, while the same mountain lion rested near one of its kills. Lawrence sat down in the open, approximately thirty-five yards away, so the cat could keep an eye on him. He again turned his back to indicate he was not any threat, after which the cat repositioned itself to nap and rest. He remained with the animal through the duration of the day, just watching from his respectable distance until shadows stretched and dusk settled like a blanket on the surrounding landscape. Then the cat rose, stretched, cleaned itself and moved to the carcass to feed, with barely a glance in his direction. In Lawrence's final interaction with the male, he dared to get quite close to the animal, purposefully testing their relationship. The lion had just dispatched a marmot near his campsite and was making short work of the animal. Lawrence approached slowly, squatted in front of the big cat, and then, after a long moment, stood and smoothly walked away. According to Lawrence, the lion, for its part, watched him closely but did not acknowledge him in any other way.

Many remain skeptical of Lawrence's narrative. He was alone in the woods, so how can we know what really happened in these encounters, or if he had any encounters of all? It would be easy to dismiss Lawrence's stories as flights of fancy, these unlikely peaceful encounters with wild mountain lions, if it weren't for all the other stories that have accumulated since.

Dr. Jim Halfpenny recounts numerous experiences during which he approached and watched mountain lions in and around Boulder, Colorado, strictly following his "100-step rule."[32] He first tried it when he was informed that a mountain lion had killed a deer in someone's backyard. He was given permission to enter the site, and he chained the kill to a tree so the big cat would be forced to feed where he could watch her. He retreated 100 steps, and sat down to wait in the open. In a near perfect echo of R. D. Lawrence's philosophy, he explained that it was important to remain where the cat could see you, rather than hidden. Halfpenny believed that no one could hide from such a master of stealth, so it was better to present oneself openly, so the cat could gauge one's intentions.

Filmmaker Jeff Hogan recalled his first trip to Torres del Paine National Park in southern Chile while he was on assignment for the 1997 televised film *Patagonia: Life at the End of the Earth*. While out shooting scenery and other wildlife, he spotted a mountain lion resting among rocky outcrops. He couldn't believe his luck. He crawled to within 100 feet of the animal, carefully trying to conceal himself, but the animal watched undisturbed and seemingly paid him absolutely no mind. After a few more relaxed encounters in which he described southern mountain lions as "totally unconcerned about me," Hogan decided to push the boundaries a bit further. The opportunity presented itself when he located an active kill site in open country. He laid down his blind and rolled out his sleeping bag some fifty yards away from the carcass, in hopes of filming the cat when it returned to feed. "I woke in the dark to the sound of bones crunching. It was magic. I rolled over and as the light began to improve I watched this female and her kitten feeding right there in front of me. Just as the light was beginning to improve more, they moved away, preened for a bit, and walked away. I just watched her leave. She was beautiful."

While on assignment for the December 2013 article "Ghost Cats," *National Geographic* photographer Drew Rush set motion-triggered cameras at active mountain lion kill sites with Panthera's Teton Cougar Project. He fondly recalls one female, called F51, who watched him several times from just fifty yards away while he set up his camera and lighting system on her kills. She watched, curious but completely passive, and then, along with her kittens, quickly returned to her kills when he departed.

There are many such stories of people approaching wild mountain lions without repercussions, and if we compare their details, the common thread may be *open country*. Open forests and other habitats in which people are easily spotted allow mountain lions to watch people approach and to assess whether they are a threat. Mountain lions, it seems, do not appreciate surprises. But this is just one man's speculation, not science. We live in an era of stories, and it is easy to gather a few to support one's speculation. Stories are winging through the air all about us all the time, invisible. They are the building blocks of social media, which is now embedded in our lives, deeper than any tick can dig. The problem we face today is differentiating fact from fiction.

The Internet is truly a wonder. It has leveled the playing field, providing equal access to information for all people, not just the privileged. The Internet spreads knowledge faster and further to more diverse audiences. Because of it, science is becoming more accessible to everyday people, and without doubt, social media is now the quickest, most-effective means of spreading information to broad audiences. People use social media to educate, inspire, and rally support or opposition for diverse causes. Because of the Internet, mountain lion facts should be common knowledge, and because there is equal access for all, people should agree on them.

Social media, however, harbors a dark underbelly. It provides all people equal opportunity to become experts on any subject they choose, and for all people to educate and motivate others to support their causes. This sounds democratic in theory, but it has created confusion for the masses too lazy to do their homework and determine whether facts floating in cyberspace originate from a reliable source. Misinformation, whether it is spread unintentionally or maliciously, is very real, and it has the potential to derail sound science, conservation, and wildlife management. People are also becoming more proficient at this sort of warfare. Here are some assertions that various mountain lion experts have presented on the Internet:

- Female mountain lions bear litters of one to five offspring every year.
- 60 percent of kittens are female.
- Because they are elusive animals and hide their kittens, 99 percent of all newborn kittens survive until adulthood.
- Kittens are spotted for three months and then become the color of adults.
- Male mountain lions weigh 180–260 pounds.
- Mountain lions kill for pleasure and abandon nine out of every ten deer they kill, without eating them.
- Mountain lions attack people who approach their food caches.
- Mountain lions do not have any natural predators.
- It is impossible for mountain lion hunters to tell male from female lions when looking at them in a tree.
- Mountain lions were never completely extirpated from New England, where they continue to breed and raise kittens in small numbers.

- If not controlled by humans, mountain lions will proliferate out of control.
- If we do not stop hunting mountain lions, we will drive them to extinction in the United States.
- Mountain lions lost their fear of people following the cessation of the bounty hunts conducted in the nineteenth and early twentieth centuries.

Every mountain lion "fact" listed above is false—every one of them.

At its best, social media is wonderful and beautiful and brings people together. At its worst, it is evidence of a growing loneliness in our culture and is a means of sowing division among communities and nations. For example, there have always been people who hunt animals and others who oppose hunting, just as there have always been those who wish to eradicate predators and those who wish to see them thrive. Today, social media is where these people share their views, as well as bully, shame, and intimidate those who think differently than themselves. So people drift toward virtual communities composed of people with similar world views and values, and with whom they feel comfortable. This sounds perfectly innocuous, but we must be careful about only interacting with those who think like us, because it distorts reality. These Internet "echo chambers," as they are now called, allow misinformation to be repeated and amplified until it becomes mainstream thought for entire groups of people. Misinformation is transformed into "alternative facts" that people believe and defend.

This book is intended to be a resource of useful information as much as a call to action. There is a lot of good found on social media today, and a lot of new information we've learned about mountain lions—information worth celebrating and spreading among friends, family, and the wider world. Linda Sweanor and her husband, Ken Logan, for example, documented mountain lion behaviors recorded after 256 close encounters initiated by researchers. Mountain lions were only aggressive on sixteen occasions, and all of these were during interactions with females with small kittens.[33] In only two of these sixteen encounters did mountain lions do anything more than hiss and growl—they charged forth from within dens where they were nursing tiny kittens, and were subsequently diverted by researchers yelling and waving their hands. Mountain lions generally fled approaching humans, or remained still, relying upon their camouflage to protect them.

Female mountain lions spend an astounding 82 percent of their lives in the company of kittens.[34] They sound like the low rumble of an engine, purring continuously to their newborn families in fortified dens that offer protection against the elements and potential predators. Females play with their kittens as they age, and close-knit families exhibit affectionate rubbing, licking, and nuzzling, even when the male kittens outgrow their mothers. Twice, Panthera's Teton Cougar Project in northwest Wyoming documented females that adopted kittens born

A rare glimpse of a female mountain lion in northwest Wyoming, nursing defenseless kittens in her den. The camera utilized invisible 940nm IR light to minimize intrusion.

to another female.[35] And only recently it has come to light that adult mountain lions are more social than previously thought—regularly interacting with their neighbors. They live in tight-knit communities defined by the territorial boundaries of resident male mountain lions.[36] Adult mountain lions share food with other mountain lions, exhibiting reciprocity with those that share with them in turn.

It's a great deal to process, all the anecdotes of people who have had peaceful encounters with mountain lions, all the new research that has begun to reveal their personalities and intimate behaviors. Collectively, it may be that we are finally beginning to understand a small part of what it means to be a mountain lion. But perhaps the greatest insights into mountain lions are occurring right now, some 6,500 miles south of the United States at the "End of the World"—in Patagonia. Torres del Paine National Park in southernmost Chile has become the epicenter of a social experiment called "mountain lion tourism," a practice that flies in the face of everything we in North America think we know about the species. The experiences of Chilean puma guides make those of Lawrence, Halfpenny, Hogan, and Rush look tame by comparison.

Rodrigo Moraga of Photo Trackers stood with a small group of photographers along one of the few roads on the ranch called Estancia Laguna Amarga, adjacent to Torres del Paine National Park. Across a small lagoon and way up the hillside, they watched the forms of running animals amidst grass clumps and globular bushes. They were far enough away that they needed binoculars to appreciate that it was a mountain lion chasing a guanaco. A guanaco is a wild llama, orange on top and white beneath, like a pronghorn. The mountain lion was a juvenile entertaining itself while its family lounged nearby. Each time the cat paused in its pursuit, the guanaco stopped, too. Then it belted out its tremulous, high-pitched alarm to alert every living creature for some distance around that a mountain lion was on the prowl, even if an incompetent one.

After several rounds of chase-the-guanaco, the cats on the ridge stood, stretched, and began their descent, collecting the energetic youngster as they passed. They were a family of four—a female and three large, one-year-old kittens. They looked like an African lion pride on the move, but on the wrong continent. The muscles in their shoulders balled and flexed above their heads as they glided downhill. Their large paws pushed backward to propel them along, flicked forward, and then rolled down, making little sound.

Moraga indicated to the others that it was time to move. He led the photographers away from the safety of their vehicles, and guided them on a course to meet the descending family. They hiked to the far end of the laguna, a green oasis of saturated earth and spongy grass. The lions arrived at the same time as the people, momentarily disappearing from view as they navigated the taller, prickly *calafate* (barberry) bushes at the base of the hill. The female, which Moraga called Sarmiento, pushed up a hare in her passage, and she quickly leapt forward to seize it in her jaws. Sarmiento was easy to distinguish from her kittens because she was orangey-brown, a familiar hue not unlike that of cats in North America. Her kittens, however, were varying shades of gray. She had two male kittens, both about her size, and one female, noticeably smaller and more timid than her brothers and mother.

Her kittens also noticed the hare, and they quickly surrounded their mother and absconded with the prize. The carcass was dismantled and consumed faster than you could say "supercalifragilisticexpialidocious." Clumps of fur clung to nearby grasses, and the severed hind feet were left behind. (The quick, casual way that hare was hunted and killed may well have revealed what North American researchers miss in foraging studies that rely upon technology instead of direct observation.)

Then the lion family emerged onstage by the open laguna before the photographers. Wild mountain lions. No radio collars, no ugly ear tags. No fencing between them and the photographers. The people were so close they could hear the big cats breathing and watch the changing colors in their rippling coats. The kittens began chasing each other in increasingly complex loops among the thickets and grass tufts. Their sinewy grace and strength were apparent in every stalking pounce, grapple, and evasion. They even tackled their mother a few times, though she tried to remain detached from the antics. Their expanding circles threatened to include the photographers as well, where they stood or squatted, riveted, about fifty meters away. Shutters clicked and whirred as the cameras recorded the secret lives of wild mountain lions. After about fifteen minutes, the running and chasing slackened, and then the family flopped down in the grass for a quick rest. Kittens nuzzled their mother and each other. Several lay atop one another in a pile.

Sarmiento clearly had a destination in mind, and after several minutes she stood to walk south. As she awaited her young troop at the far end of the laguna, she looked back over one shoulder, barking the mountain lion's unique call—a single syllable, hoarse yet high-pitched bark. When that failed to pull her kittens to her, she wandered back to collect them, and then the family continued southwards toward the lake.

It was just a family of mountain lions being mountain lions—the sort of thing researchers in North America never get to see, though many would give their right arm to do so. Most North American researchers spend their entire careers trying to learn something about mountain lions from the blips and bleeps of handheld receivers, rather than from any opportunity to actually observe the animals, except when they are trapped, or cornered by hounds. In Patagonia, however, mountain lions are like bison or bears in Yellowstone National Park. They can be watched. Their every rolling step, their every flick of the tail, is an education—an undeniable truth forcefully displacing our construct of what

Two of Sarmiento's one-year-old kittens chasing each other in an endless game of cat-and-mouse in southern Patagonia.

mountain lions must be like from our 200 years of chasing them with hounds and 60 years of fitting them with radio collars.

Sarmiento and her kittens, for their part, were obviously comfortable with being observed. They weren't running away from the photographers, or stalking them slowly in planned attacks. They merely scrutinized the photographers as they strolled by, fifteen meters away. Their gaze passed over each person, a tangible pressure the photographers couldn't help but notice. It was as if the lions weighed and measured their observers in a mere glance.

Sarmiento and her kittens eventually dropped to the lakeshore itself, where they drank from the shallow waters; the scene looked like stock footage of captive mountain lions in Marty Stouffer's *Wild America*. It was like being on an African safari, minus the vehicles and rifles. Then Sarmiento and each of her kittens filed past the photographers, who were squatting on the lakeshore about thirty meters from where they were drinking. They moved past the photographers cautiously, but intentionally, within arm's reach of several of them.

Clearly some photographers hadn't been told not to lock eyes with a predator in case that would trigger an attack, or maybe they couldn't help themselves. Several maintained their composure, and their cameras whirred. Others were confounded by the proximity of the large cats and the limitations of their telephoto lenses to focus on anything so close. They sat back and stared, lost in the unnerving intensity of Sarmiento's stare. She was not threatening in the least, but her fluid, rolling grace exuded strength and confidence. Her kittens followed her one after another. The photographers sat quietly, transfixed by the afterimages that played across their imaginations.

Across the water to the south, Rodrigo pointed out the place where, two decades previous, a mountain lion had killed a local fisherman angling for brown trout in a rivulet emptying into the lake. Much more recently, another mountain lion had attacked a bicyclist in the middle of the day in central Chilean Patagonia, injuring his arm. Nevertheless, these resilient people continue to seek the company of mountain lions. And more, they trust mountain lions enough to spend time with them, armed only with cameras and tripods.

After their unexpected encounter with M58, Patrick and Michelle drove their sleds deeper into the National Forest to visit more areas where mountain lions had rested and killed prey. They gathered data essential to revealing the little-known ecology of mountain lions in northwest Wyoming. On their way back to the office, Patrick broke the silence that had settled over them following a long, physical day in the bitter cold.

"Did it look like he was going to jump on me?" he asked, somewhat sheepishly. Michelle laughed, not because of the question but rather

A wild adult female mountain lion pauses to look for her kittens, which have fallen behind.

because he'd obviously been pondering the possibility all day without bringing it up. She assured Patrick that M58 had never shown any sign that he might attack him. She related how M58 had looked only at her, and hadn't acknowledged Patrick at all until he moved.

With this explanation, Patrick visibly relaxed and smiled. Even researchers sometimes doubt their understanding of mountain lions.

The question remains, however: Which are they, demon or cur? Perhaps mountain lions are really both—timid and ferocious—and in this reconciliation is the true animal. As Chris Bolgiano wrote in her book *Mountain Lion: An Unnatural History of People and Pumas,* "Mountain lions are as fierce as bloodshed but gentle as a mother with kittens, as strong as death yet vulnerable to accidents and age." But perhaps, too, they are neither. "Demon" and "cur" are both constructs of human imagination, lingering reminders of an era when predator eradication was at the forefront of every form of wildlife management. These constructs persist today because they are maintained by our fear of animals with big teeth, and because we haven't developed new constructs. So perhaps it is not mountain lions that need to change, but humans.

The people who live in and adjacent to Torres del Paine National Park in southern Patagonia are trying something different. They are trying to live *with* mountain lions, not in spite of them. It is a bold and beautiful model, but risky, too. Should it fail, people, mountain lions, and local communities built on tourism will suffer. Nevertheless, people's experiences with mountain lions in Patagonia should challenge how we perceive these animals.

As we observe and study mountain lions more, our understanding of these magnificent creatures continues to evolve, providing new fodder with which to imagine mountain lions as something more than snarling killing machines. Evidence suggests that mountain lions are intelligent, curious animals absolutely ruled by caution. Their family life is filled with affection and play, and as adults they exhibit strict social norms that regulate their social interactions with other mountain lions. They are capable and efficient hunters, adept at disappearing, and exceptionally strong for their size. They are generally peaceful, even when provoked, and almost always choose flight over fight. They also appear to be tolerant of respectful intrusion and quite willing to peacefully coexist with humans—if we choose to allow it.

CHAPTER 2
Staying Safe in Lion Country

It was mid-September 1993, and the Kowalskis were among the many families enjoying Paso Picacho Campground in Cuyamaca Rancho State Park in eastern San Diego County in Southern California. The park had just reopened after a brief closure due to what officials felt was abnormal and potentially threatening mountain lion activity. Lisa Kowalski, just ten years old, was playing cards with her father and brother when an emaciated female mountain lion materialized out of the brush alongside their campsite.

"Daddy yelled for me to stand still and then I looked around and saw the mountain lion and I screamed," Lisa later recounted. "The mountain lion sniffed me and then bit me real hard. I screamed again and it let me go."[1]

Mr. Kowalski acted brilliantly and yelled to his wife to release their lab-terrier mix, which was tied to their RV at the time. The cat and dog engaged in tooth and claw, and the hound succeeded in driving off the unhealthy cat. The sickly lion retreated into nearby brush, where it remained until two park rangers arrived and dispatched it with shotguns. Both Lisa and her dog were treated for wounds—luckily both were little hurt during the incident.

The lion also matched the description of the animal that had caused the park closure several weeks earlier. Just one month before Lisa Kowalski was bitten, well-respected mountain lion researcher Dr. Paul Beier said, "I think most would agree with me that cougars are no more dangerous to humans than breathing the Southern California air."[2] His underlying message was clear: mountain lions are hardly dangerous, the chances of being attacked are slim to none, and instead we should worry about the pollution we breathe in, which is definitely killing us, even if slowly. In retrospect, Beier's message was reflective of the calm before the storm, a statement that few would have contested in August of 1993. What Beier did not know at the time—what he could not know—was what was soon to come. In a cruel stroke of fate, California and the West were immediately struck thereafter with the worst streak of mountain lion attacks in recorded history.

On April 23, 1994, Barbara Schoener, a marathoner, took to the American River Canyon trail in Auburn State Recreation Area, which lies hundreds of miles to the north of San Diego County where little Lisa was attacked.[3] Barbara had run those trails many times before. She was attacked by a mountain lion from behind. She managed to stand and run again, and was attacked a second time. Barbara was the first person to be killed by a mountain lion in the state of California since 1909, when Isola Kennedy died from rabies after battling a lion that sprang on a young boy in her church group. Isola was buried in Morgan Hill, California, under an epitaph that read: "Sacrificed her life battling a lion to save some small boys."[4] Eighty-five years later, Barbara was survived by her husband and two children. The suspected lion that killed her was tracked down and slain.

Just four months later in Northern California, a mountain lion attacked a dog and a woman, Kathleen Strehl, while they were on a camping trip with a small group of others. The two men in the group managed to grab the lion after it knocked Kathleen down, and to hold it down while her friend Robin Winslow stabbed the animal with a large kitchen knife. Troy Winslow lost his thumb because he stuck it in the cat's mouth, triggering a bite reflex.

"I thought he'd killed my dog," Robin later explained. "I was stabbing him like he killed my dog."[5] She managed to kill the mountain lion, dispatching the animal before it inflicted much harm other than the loss of a thumb and a scratch on Kathleen Strehl. (Robin's dog, it turns out, was fine and had run away, just as any sensible creature would do.) But so fresh was the incident on the heels of Barbara Schoener's death and the attack on Lisa Kowalski that the people of California were beginning to stir. In fact, Assemblyman David Knowles of Placerville, where Barbara Schoener had resided, introduced a bill in 1994 to repeal the twenty-two-year-old ban on mountain lion hunting in California, a move many hunters were quick to support.[6]

"As predicted by the sportsmen of California," stated Dan Heal, then chairman of the California Sportsmen's Task Force, and a vehement supporter of the proposed legislation, ". . . unwarranted protection of a predatory animal—the mountain lion—has come full circle with the death of Barbara Schoener."[7] The bill ultimately failed.

The cougar conundrum had been simmering for decades, unbeknownst to most Americans, but during the 1990s it began to boil. People in the West were suddenly faced with a new reality—one in which they lived alongside a large predator. For some, it was as if nightmares had come to life, while for others it was evidence of recovering ecosystems and therefore a reason for celebration. Approximately thirty-five mountain lion attacks occurred in the United States and Canada during the 1990s; six of them were fatal. (There was a potential seventh person killed as well, but it remains an unverified report.)

Mountain lions are a threat to human safety, even if that threat is miniscule. Let's agree on the fact of this threat as a starting point for discussing the implications for those of us who want to live with mountain lions. State wildlife agencies immediately hunt down any lion that attacks people, and for the most part they are successful in catching and dispatching dangerous lions. Trained hounds are generally the means of doing so; sometimes the hounds are owned by state agency personnel, but much more often they are trained by personnel working for Wildlife Services, the federal program responsible for killing most wildlife in our country, or by private citizens.

Removing culpable lions that attack people is part of the solution to the cougar conundrum, and it's part of a good strategy for living with mountain lions in a modern world. State agencies cannot afford to allow a mountain lion to run free after it attacks a person, hoping it won't do so again—especially with our growing population and litigious culture. For example, the family of a five-year-old girl who was attacked and disfigured by a mountain lion in Southern California in 1986 successfully sued Orange County for negligence and failing to warn the public about known dangers in county parks.[8] One consequence, however, was that Orange County closed the entire of Ronald W. Caspers Wilderness Park to minors until 1995, thereby halting any use by families. Similarly, the parents of a mountain biker killed by a lion in Southern California in 2004 sued Orange County for negligence, but they dropped the suit following pressure by local and national bike associations concerned over backlash that might limit mountain-bike access to public lands. State agencies will continue—and should continue—to hunt perpetrators in order to keep people safe. Our job then must be to reduce the need for lethal state-agency interventions if we want to encourage peaceful coexistence with mountain lions.

Mountain lion attacks on people in the United States and Canada have never been frequent, but they began to increase in the 1970s. The number of attacks peaked in the 1990s and have since declined, though they continue to occur more often than they did before the 1990s. Mountain lions aren't changing and becoming more aggressive. Most agree that the increase in attacks is just a numbers game. Mountain lion attacks increased following the increase in the number of mountain lions on the landscape, the increase in the number of people in the United States and Canada, and the increase in the number of people using wild lands to bird-watch, hike, backpack, ride horses, fly-fish, hunt elk, and on and on. Recreating outside is more popular than ever, and there are simply more mountain lions and more people than there were in the first half of the twentieth century. If you take into account the number of people, for example, and quantify the number of attacks per person in the United States, the 1990s only experienced roughly double the number of attacks that it did in the 1950s, which was one of the worst decades in the years previous to the end of bounty hunting. Over this same time period, the human population nearly doubled as well. If we knew the changing number of mountain lions over the same time period and were able to estimate the number of attacks per mountain lion per person each year, we'd likely find that mountain lion attacks haven't risen at all—they almost certainly occur at the same incredibly low rates that they always have. Mountain lion attacks are just sensational.

This is not meant to minimize the risks associated with living with mountain lions and other large predators, risks that are very real, even if small. Chris Ingraham compiled deaths recorded in the Center for Disease Control databases on animal-caused mortalities that occurred in the United States across thirteen years from 2001 to 2013. On average, sharks, alligators, and bears each killed one person per year. Venomous snakes killed six people per year and venomous spiders killed seven people per year. The really dangerous animals, however, were common domesticated species. Cows killed 20 people per year and dogs killed 28 people per year. Over the 118 years between 1890 and 2008, 21 to 29

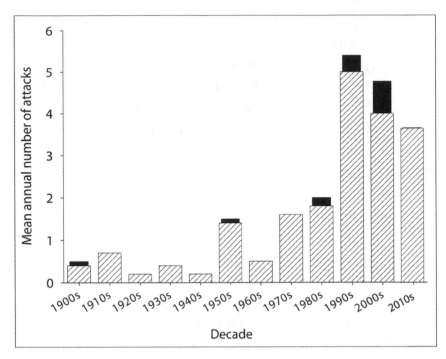

The average annual number of mountain lion attacks on humans for ten-year periods going back to 1900. This graphic was originally published in Mattson, Logan, and Sweanor, "Factors Governing Risk of Cougar Attacks on Humans," 141, and is modified here to include data for 2010–19. (Used with permission of D. Mattson.)

people were killed by mountain lions—the larger number includes unverified but possible incidents.[9] Over 118 years, mountain lions killed approximately equal numbers of people to what cows and dogs kill in the United States every year. Deer kill up to five people and injure hundreds each year in the United States in physical attacks—but, there are never media alerts warning people that a deer has been spotted lingering in a city park or wandering through someone's backyard. Between 1978 and 1995, 37 people were killed by vending machines, which is nearly double the number killed by mountain lions over a time period six times longer.[10] It's a dangerous world out there, but mountain lions pose little risk when compared to all the other dangers.

"We are not on the menu," said Doug Updike, the California state biologist responsible for mountain lion research and monitoring during the 1990s. "If a lion had any desire to catch and eat people, we would see literally hundreds of people dying every day."[11] Instead of focusing on the potential dangers of mountain lions, many researchers have speculated about why mountain lion attacks on people are so very rare, given the animals' capabilities and the opportunities mountain lions have for hunting people in wooded areas where recreation is popular. People weigh about the same as a deer, and thus should be the perfect size for mountain lions. Some speculate that lions simply never learned that people are prey. Other researchers believe that humans are just the wrong shape, standing tall like a tree trunk rather than presenting the horizontal back and curved neck of a deer or an elk. But mountain lions' reluctance to attack humans may also be something peculiar to lions, something we have yet to understand. The closest relative of the mountain lion among the world's other big cats is the cheetah, a timid species for which there exists not a single documented attack on a person in the wild. Both cheetahs and mountain lions are uniquely caught in the middle—large cats that live with larger, stronger, fiercer predators than themselves. Both are diffident as compared to other large predators, and choose flight over fight unless no option to flee is left to them.

Our fear of large predators, however, is rarely rational. For example, 55 percent of 247 survey respondents from Alberta reported that they believed mountain lions posed an equal risk to their survival as did driving a car. In fact, only one person has been killed by a mountain lion in Alberta over the last 100 years, while 400 people died in vehicles each year for the five years leading up to the study.[12] Numbers and statistics also do nothing to mitigate people's fears of mountain lions. Contrary to expectation, new research conducted by social scientists suggests that bludgeoning people with statistics on the rarity of attacks may even *increase* fear of predators rather than reduce it. Media, of course, feeds our fears as well. Too many articles about mountain lions in the media today are local alerts and warnings that mountain lions have been seen in neighborhoods, near schools, or in recreation areas, and so we should keep a watchful eye on children and pets. Thirteen authorities on mountain lions wrote the following: "Arguments for decreasing cougar density often focus on scenarios of cougars lurking near homes and settlements. . . . This argument may be nothing more than a rhetorical device to promote regional hunting."[13] In other words, fear of mountain lions is being weaponized in order to push lethal control of predators.

California is unique in the West in that it does not support any form of legal mountain lion hunting for the general public. This fact is particularly intriguing, given that California was once one of the most aggressive states in trying to eradicate mountain lions from natural landscapes. Between 1907 and 1963, the state of California paid out 12,461 bounties for mountain lion carcasses.[14] The bounty system in California was curtailed in 1962, but the killing continued, unmonitored. In 1969, California opened its first managed mountain lion hunt, during which people who wished to hunt mountain lions were required to purchase a license to do so.

Ironically, the state's new mountain lion hunt coincided with a request from an old deer hunter to protect mountain lions. Lowell Dunn wrote to his state assembly representative in Napa County, John Dunlap, to advocate for the species. Dunn said that he thought something needed to be done to protect the big tawny cats, given that he had only seen them twice in thirty years of deer hunting.[15] He hoped he'd get at least one more chance to glimpse one before he died. Dunlap took up Dunn's cause with unexpected fervor and flair, confronting

Assemblyman John Dunlap, accompanied by a lion ambassador, announces his proposal for a mountain lion hunting moratorium in Sacramento in 1971. (Unlisted photographer, from the front cover of the *San Bernardino Sun*, February 26, 1971.)

hunters and cattlemen who opposed the legislation. "Anyone with fifty cents for a tag and enough money to buy shells for his gun can hunt the mountain lion today," he proclaimed. "There is no hope of saving these animals unless we move now."[16] He presented the bill on Capitol Hill, flanked by two tame mountain lions called Huntley and Brinkley, that were lying across his desk. Dunlap's moratorium was signed by then Governor Ronald Reagan.

Dunlap's moratorium went into effect in February 1972, at the close of the state's second legal mountain lion hunting season. In the state's two brief hunting seasons, the state earned revenues from the sale of 4,953 hunting licenses, and hunters legally killed 118 mountain lions. Hunting moratoriums were renewed several times and mountain lions experienced uncontested protection until 1986, when, in preparation for reinstating legal hunting, the California Department of Fish and Game reclassified mountain lions as a game species. Mountain lion advocates challenged the change in court during 1987 and 1988. Subsequently, the Mountain Lion Preservation Foundation, now known as the Mountain Lion Foundation, and others initiated a successful campaign to introduce a ballot initiative called the California Wildlife Protection Act, or Proposition 117, which passed by a narrow margin in 1990. Prop 117 simultaneously undermined state wildlife-management plans and permanently designated the mountain lion as "a specially protected mammal." The passage of Prop 117 was met with disbelief, if not scorn, by many people living in other Western states. Lion hunters and their advocates argued that California's agencies had foolishly left the state's people unprotected. The string of attacks that began with Lisa Kowalski served to renew debates over protecting mountain lions.

Barbara Schoener's horrific death in 1994 was enough to make 1994 an unusual and terrible year for lion–human relations everywhere, but even the follow up attack in August that resulted in Kathleen Strehl's injury wasn't the end of it. On December 10, 1994, fifty-eight-year-old Iris Kenna was walking along a dry, lonely trail lined with heavy scrub back in Cuyamaca Rancho State Park, near the junction of two fire roads, Lookout and Azalea Springs. She was attacked and killed by a young male mountain lion weighing 116 pounds. Cuyamaca Rancho State Park closed a second time in as many years due to mountain lions, while professionals tracked and killed the animal.[17] In early 1995, a cyclist was attacked north of San Diego; he escaped with minor injuries after pelting the lion with rocks and defending himself with his bicycle. Between 1993 and 2000, visitors to Cuyamaca Rancho State Park reported 201 mountain lion sightings, an incredible number given the secretive nature of the beast. During the same time period, Park officials recorded sixteen instances of mountain lion behavior regarded as threatening, and wildlife officials euthanized nine lions for reasons of human safety.[18]

Stimulated by the blood spilled during the rapid succession of attacks, the media were in a feeding frenzy, and increasingly media articles presented mountain lions in a negative light.[19] The stage was set for a more concerted assault on California mountain lion protections and Proposition 117. It came in the form of Proposition 197, legislation authored by California state senator Tim Leslie, and widely supported by livestock associations, various sporting associations, and the National Rifle Association.

"Mountain lions and humans are on a collision course," explained Senator Leslie. "I only hope that we can reverse this trend before those incidents lead to more attacks or deaths."[20] In the end, Proposition 197 was voted down 58 percent to 42 percent, which was a larger margin than in the original passage of Proposition 117 in 1990.

Even in the face of a string of mountain lion attacks, the people of California voted to continue to protect mountain lions from hunting. Some folks thought Californians were absolutely bonkers to allow a large predator to live in the state uncontested. Referring to a particular attack then in the news, a California sportsman wrote that "The mountain lion attack itself would not have happened in the first place if mountain lions were still hunted, or at least chased, with trained hounds," and "over the decades since mountain lion hunting was stopped, lions have lost their fear of man."[21]

Numerous houndsmen, hunting advocates, and state wildlife personnel believe with absolute certainty that hunting mountain lions makes them afraid of people and therefore increases human safety.[22] Hunting, then, is perhaps one strategy to reduce the risks posed by mountain lions to humans and a way to facilitate human–mountain lion coexistence.

The idea that hunting mountain lions increases human safety is built upon two related but distinct suppositions: first, that hunting mountain lions makes them afraid of people; second, that hunting mountain lions with hounds increases human safety. Ultimately, if these assumptions are true, there should be fewer attacks in areas with greater hunting pressure, and more frequent attacks in areas without hunting, like California.

Let's begin with fear. Mountain lions, by their very nature, are timid and reserved. Historical accounts stretching back to the earliest written accounts by European explorers emphasize this fact. Early hunters went as far as to express disappointment over the timid behavior of cornered mountain lions and the ease with which they could be shot. Remember that President Teddy Roosevelt said, "It is absolutely safe to walk up to within ten yards of a cougar at bay, whether wounded or unwounded, and to shoot it at leisure."[23]

Mountain lions appear to fear people regardless of hunting pressure. During research conducted in California, where mountain lion hunting is illegal, Dr. Justine Smith found that mountain lions confronted by human voices immediately abandoned hard-won food they'd acquired.[24] Smith created a playback experiment in which she and her team placed recorders at active mountain lion kill sites. Vocalizations of Pacific tree frogs elicited no response from the big cats—but recordings of talk show hosts Rachel Maddow and Rush Limbaugh were another story altogether. Cats fled the scene in 83 percent of cases, and in half of cases they never returned to feed from the kills they needed to sustain themselves. When they did return, they took longer to do so after hearing people and they ate less of the carcass. Smith concluded, "Pumas are nonpartisan in their hatred of American politics" but, more importantly, mountain lions fear people to such an extent that they will abandon food and expend additional energy to hunt again rather than risk an encounter with people.[25] And this was mountain lion behavior in the absence of legal hunting.

Speaking more directly to the question of hunting and fear, we lack clear evidence to determine whether chasing cats with hounds makes them afraid of people. "Hazing" is a common tactic used on bears that become habituated to feeding in campgrounds. They're shot with rubber bullets and chased away by hounds, and this generally works to scare bears back into areas where they are less likely to get into mischief with people's food or refuse. Sometimes, though, individual bears are not dissuaded by hazing, and then they may be captured and removed to an area deemed safe for them to wander. As a last resort, "problem bears" are euthanized. Such hazing tactics have rarely been tried with mountain lions, but based upon the success of hazing in reducing negative human–bear interactions, it stands to reason that it might work on individual mountain lions as well. Rich Beausoleil, of Washington State Department of Fish and Wildlife, has tried it often, and is skeptical that hazing with dogs changes mountain lion behavior.

In the end, we want to know whether hunting actually increases human safety, and in this regard we do have a clear answer. According to the thirteen mountain lion authorities who authored the Cougar Management Guidelines, "Sport hunting has not been shown to reduce risk of attack on humans." They go on to write, "There is no scientific evidence that sport hunting achieves this goal. In rare cases where a cougar exhibits dangerous behavior and needs to be removed, this job is best done by a professional to expeditiously track and kill the individual cougar, rather than via sport hunting." It doesn't get any plainer that that.

In fact, California, where mountain lion hunting has been banned for decades, experiences fewer human–mountain lion incidents than any of the other nine Western states that permit hunting and collect similar data.[26] California also experiences among the least frequent mountain lion attacks among Western states, when measured in attacks per million people. In contrast, mountain lion attacks, especially fatal ones, have occurred more often on Vancouver Island, off the coast of British Columbia. No one really knows why cats there have attacked people more often, but it is an area where lions experience heavy hunting pressure and long hunting seasons.

Therefore, there is no support for the idea that hunting increases human safety. Mountain lion hunting generally occurs away from people and away from the very areas where people and mountain lions are more likely to interact.

There is, however, a cold logic to the idea that reducing the number of mountain lions should reduce the probability that anyone might meet one, so therefore hunting should increase human safety. Nevertheless, this logic is flawed because it makes two false assumptions. First, it assumes that all mountain lions are the same and behave in the same ways, which they certainly are not and do not. Second, it assumes that every mountain lion removed from an area results in a permanent reduction in the number of mountain lions in the mountain lion population. The effects of hunting on mountain lion populations are discussed more in chapter 5, but for now, suffice it to say that populations don't respond so simply. In fact, hunting pressure often removes resident males and results in an influx of young mountain lions trying to fill the void. As it is young mountain lions that are more likely to threaten people and livestock, this may be problematic—but we're getting ahead of ourselves.

The Montana Department of Fish, Wildlife & Parks manages mountain lions with what they call a "zone system," in which they attempt to maintain a mountain lion–free buffer around human communities.

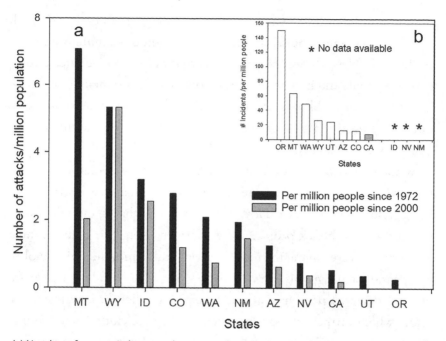

(a) Number of mountain lion attacks on people per one million people in ten Western states with mountain lion hunting, and California, which does not permit lion hunting. (b) Number of mountain lion incidents, defined as the sum of attacks, threatening encounters, and pet and livestock depredations, per one million people in eight Western states with mountain lion hunting, and California. Note that mountain lion hunting does not appear to improve these metrics when compared with California, a Western state with a healthy mountain lion population but no hunting. (Originally published in Laundré and Papouchis, "The Elephant in the room," Fig. 5. Used with permission of J. Laundré.)

Mountain lions can pass through these areas, but state officials remove any that linger and appear to be establishing territories near people. The zone strategy attempts to reduce the chances that mountain lions might become accustomed to people as neighbors. The concern is that habituated animals may lose their fear of people and eventually do something they ordinarily wouldn't, like attack a poodle or worse, a child. Nevertheless, we lack research to test whether zone management increases human safety or is just a placebo meant to placate nervous constituents in communities with mountain lions. Again, cold logic would dictate that if the state agency remains vigilant in killing mountain lions, and is able to maintain a buffer around people without mountain lions, it could work. Nevertheless, history shows that most negative interactions between people and mountain lions involve young mountain lions lacking their own established territories. Therefore, zone management is unlikely to protect people completely, and worse, may be killing mountain lions unnecessarily.

While there exists no evidence that hunting increases human safety, numerous researchers have instead written warnings that heavy hunting pressure may in fact *increase* the chances that mountain lions will threaten or attack people. Dave Mattson, Ken Logan, and Linda Sweanor, who are among the most knowledgeable mountain lion biologists, conducted rigorous analyses of previous mountain lion attacks in order to decipher patterns that could help us predict the circumstances under which further attacks might occur. They concluded that "Young cougars in poor condition are more likely than other cougars to threaten people," and that, because hunting often results in an increase in the number of young mountain lions running about on the landscape, "heavy localized hunting of older cougars could increase rather than reduce exposure of people to close-threatening encounters with cougars."

On February 4, 2019, a mountain lion attacked Travis Kauffman while he was out for a jog on the West Ridge Trail in the Horsetooth Mountain Open Space in the foothills of the Rocky Mountains near Fort Collins, Colorado. Travis fought back, successfully choking and killing the animal, and then continued down the trail on his own before he met up with other folks who helped him get off the mountain and receive medical attention. The hospital called the Colorado Parks and Wildlife Department (CPW), as is routine in wild-animal attacks. CPW staff arrived at the scene of the attack at about 5:30 p.m. to corroborate the story and collect the carcass of the mountain lion in order to perform a necropsy. Necropsies of lions that attack people are standard protocol for assessing the animal's health as a possible explanation in what caused the attack in the first place.

Headlines the next morning were rife with drama: "Colorado Jogger Strangles Mountain Lion after Attack," "Colorado Jogger Strangles Mountain Lion to Death with Bare Hands after Attack on Trail near Denver," "Colorado Jogger Chokes Mountain Lion to Death, Then Drives Himself to Hospital after Being Attacked," and "A Runner Strangled a Mountain Lion after It Attacked Him." The articles recounted the epic battle, during which Travis succeeded in wrestling and choking an eighty-pound mountain lion to death with his bare hands. By the second day following the attack, he was legend.

The articles, unfortunately, were wrong about a lot of things. CPW decided to trickle out information in an attempt to control the public response to the incident, but managed only to contribute to the misinformation being shared. Travis himself experienced severe lacerations on his hands, arms, and face, but luckily none of his wounds were life-threatening. Travis had done everything right—when he found himself the victim of a mountain lion attack, he fought back and he survived. He even avoided the media for ten days. Then CPW arranged a Valentine's Day press conference, during which they recounted events up to that point and introduced Travis to the media to tell his story. Again, they decided to disclose minimal information about the mountain lion that attacked Travis. They left clues, however, for those who knew enough about mountain lion natural history to determine that it was no eighty-pound beast that had attacked Travis, but rather a kitten, and likely an orphaned one at that.

As Travis jogged beneath rock turrets and outcroppings, he heard something moving on pine needles and downed branches behind him.

He looked over his shoulder to see what it was. "[I] was pretty bummed out to see a mountain lion running after me," he said during the press conference, with what we soon learned was his characteristic under-statement and humor. The mountain lion, as it ends up, was four to five months old and weighed just thirty-five to forty pounds. It jumped up for his head. Travis raised his hands to defend himself and the little lion latched onto Travis's wrist, lashing out with claws at his face and refusing to let go. In a frightening, adrenaline-filled encounter, Travis managed to wrestle it into a position where he could pin it beneath him and step on its neck to choke it out. When the cat finally released its bite, Travis stumbled away down the mountain until he encountered others who assisted him. Later he joked, "If it was Chuck Norris, he would have come out of it without a scratch and then put [the cat] on his shoulder." Travis, in fact, was a bloody mess.

A nearly five-month-old mountain lion kitten, which is the age of the mountain lion that attacked and was subsequently killed by jogger Travis Kauffman in Colorado in February 2019.

The CPW and the regional parks shut down the area immediately. They placed motion-triggered cameras across the landscape in an effort to capture images of other mountain lions in the area. They never saw signs of an adult female, but they did catch images of two other kittens of the same age. They set box traps with bait and were able to catch both kittens, which they believe were siblings to the one that had attacked Travis. Both were described as "hungry," and CPW announced as early as the day of Travis's press conference that they planned to rehabilitate and release the kittens into the wild when they were ready. That was the first indication that the mountain lions were very young, as officials would never try to rehabilitate an actual eighty-pound mountain lion even tenuously involved in a human attack. Further, CPW reported that once Travis had killed the lion, its siblings had immediately fed upon its body—this, too, is characteristic behavior of orphaned mountain lion kittens desperate for sustenance. Finally, the cat that attacked Travis also had vegetation in its stomach. Mountain lions are obligate carnivores and only rarely eat vegetation to help with digestion; when they eat grass they tend to vomit it up almost immediately rather than keep it down and try to digest it.

Typically, when a mountain lion attack occurs in the United States or Canada, everyone wants to know about the animal—was it starving? Did it have rabies? Was it compromised in some way? Perhaps to help rationalize and quell any lingering fear, everyone wants to know *why* a mountain lion would attack a person. Yet even months after the horrific attack on Travis, long after CPW had finally used the words "orphaned kitten" in interviews, and the media frenzy was finally dying down, no one asked how the mountain lion that attacked Travis had become orphaned in the first place. The attack occurred in February in an area with legal hunting—and by far the most common cause of death for mountain lions in winter throughout the West is legal hunting. Could a female have been killed, abandoning three kittens that then became dangerous to Travis? When CPW was finally contacted three months after the attack to request information about female mountain lions that had been killed in the area, they were uncooperative and would not release the information.[27]

An investigative reporter named Rico Moore submitted a Freedom of Information Act request to CPW many months after the event. He discovered e-mails betraying CPW's internal communications discouraging their staff from discussing the fate of the mother of the kittens with the media.[28] They did not want to encourage discussions that might lead journalists to speculate about the potential links between mountain lion hunting and mountain lion attacks on humans. In its outreach surrounding the event, CPW instead emphasized that the attack could have "had a very different outcome" if it had been an adult mountain lion, implying that then Travis may have been more seriously injured, or worse, he may have been killed. What they should have emphasized is that it would have been different if Travis had been a six-year-old boy instead of an adult. Then the parents of America would have risen up and demanded answers regarding the potential link between mountain lion hunting and attacks on people. Everyone agrees that even the smallest risk to America's children is unacceptable. Luckily for CPW, however, another attack drew media attention away from what they appeared to be trying to ignore.

On March 31, a young mountain lion attacked a seven-year-old boy in his backyard in Lake Cowichan on Vancouver Island in British Columbia. Chelsea Bromley raced downstairs toward the terrible shrieks of her son, Zach. Seeing the lion holding Zach by the arm, she grabbed its jaws and pried them open to free her son. The cat ran off into nearby cover. British Columbia conservation officers caught and killed the cat responsible, as well as a second cat located in the same area and thought to be sibling to the first. Wildlife officials quickly announced that the offending lions were orphaned kittens: "The cougars were both very emaciated, very skinny, and in poor overall condition, very light for their size," reported Conservation Officer Ben York.[29] They did not speculate how they might have been orphaned, even though the mountain lion hunting season, stretching from September 10 through June 15, was still underway.

We know that hungry young mountain lions lacking support from their mothers are the most dangerous contingent of mountain lion populations. We also know that heavy mountain lion hunting increases the

A female cleans and bonds with her seven-month-old kitten. Kittens are rarely self-sufficient before one year of age and will likely die if orphaned before then.

number of young lions on the landscape in two ways: first, when a territorial male is taken, youngsters flood in to fill the gap; and second, the hunting of females sometimes unintentionally results in orphaned kittens, which have little chance of survival on their own until after they are at least one year of age.

One might argue that wildlife managers in charge of mountain lion hunting should be working to reduce the number of unhealthy, young animals among lion populations. Specifically, managers could slow resident male turnover by reducing hunting pressure in general, and they could reduce the chances that kittens will be orphaned by reducing hunting pressure on female mountain lions. Unfortunately, the reality is that females often travel separately from their kittens, making it impossible to protect lion family groups completely without banning the hunting of females entirely. Most managers, however, have not responded to the warnings posed by numerous respected scientists about the potential links between heavy hunting pressure and threats to human safety. Instead, the management of mountain lion hunting remains heavily influenced by the very same special interest groups that pushed for predator eradication a century ago.

If hunting mountain lions does not increase our safety, what can we do to reduce the chances that a rare encounter with a mountain lion will turn terrifying? Luckily, the best strategy for increasing one's personal safety is what it has always been: making smart decisions. Just as savvy city slickers know which streets to avoid when walking alone or parking a car on at night, there are ways to minimize the risks associated with living with lions. Researchers, for example, have studied the circumstances surrounding attacks in order to elucidate patterns that can inform us about the conditions under which a potentially magical encounter with an elusive animal may turn dangerous.

Based upon 386 threatening encounters and physical attacks over 120 years, Mattson, Logan, and Sweanor found that people who did not act aggressively when confronted by a lion, or who did not have the opportunity to do so, were more likely to be attacked.[30] They found that groups with small children that encountered mountain lions were 6.4 times more likely to be attacked than groups of adults, and groups composed only of children that encountered a lion were fourteen times more likely to be attacked. People who were active (running or biking) or moving erratically like a wounded animal were also more likely to be attacked than those engaged in sedentary activities. Dogs were a mixed bag. The likelihood of negative encounters with mountain lions is higher for dog owners at night near their homes, because dog owners tend to let their dogs outside to pee. Should those dogs be unfortunate enough to be attacked by a cat, the owners might jump into the fray to protect their pets, and they might be injured. Away from home and on the trail, people accompanied by dogs are also slightly more likely to have a negative encounter with a mountain lion—in general, small to medium-size dogs are an attractant to mountain lions, not a deterrent, though a larger dog, one prone to chasing cats, will have the upper hand. Female mountain lions are more likely to attack than male mountain lions, and younger mountain lions are more prone to attack than adults. When they do attack, however, adult mountain lions are more likely to kill a person than a younger animal.

Remember that driving a car is *much* more dangerous than walking in lion country, and carries a much higher probability of injury of death—yet we put on our seatbelts and manage to drive vehicles all the time. In the same way, we need to assess and understand the risks of living with mountain lions and make smart decisions (like putting on our seatbelts)

Recreating in big wild lands, especially public lands, is one of the great joys of living in the USA. Make good decisions and follow common-sense advice, and your time in the woods will always be enjoyable. (Artwork by Charlie Layton.)

in order to reduce the possibility that we will have a negative interaction with one.

Here are a few common-sense practices that will reduce the chances you'll encounter a mountain lion. First, mountain lions are most active in the hours before and after dawn and dusk, and to a lesser degree at night. In areas with more human activity, they tend to become more nocturnal, but exceptions abound and do not apply to every mountain lion. You may encounter one at any time of day. But, as a general rule of thumb, start your hike, bike, or jog at least a few hours after sunrise and be sure to end it a few hours before sunset in order to minimize your chances of meeting a lion.

Hike in groups to increase your safety. Keep small children younger than six years old in sight at all times—don't let them run so far ahead that you can't keep an eye on what's happening. Children older than six are tougher to keep tabs on, of course, but try if you can. Educate your children—don't make them afraid of mountain lions, but do make them respect the power of mountain lions. Explain what children should do if they encounter one. Carry bear mace (pepper spray) or an air horn (in windy conditions), and keep it where you can reach it—it's no help at all if it's buried in your pack when a bear or mountain lion drops onto the trail in front of you.

Do not use headphones or ear buds that block out natural sounds while recreating in the woods, especially if you are alone. Do not wear hoods or hats that obscure your view of your surroundings. There is nothing more foolish than jogging or mountain biking in mountain lion habitat at dusk or dawn, while wearing headphones and a hoodie, so that you can't hear twigs snapping or see animals closing in. This is good advice for avoiding bad encounters with any predator, even the human kind.

If you do see a mountain lion, the best general advice is to enjoy the show and count yourself lucky. Most sightings are brief, as a cat crosses a trail or road, or as the cat flees from you as fast as it can run. But in cases in which the cat hasn't seen or become aware of you, there is no cause for alarm. Allow it to go about its business and just watch and enjoy. Group together if you are with others. Call children or dogs in close to you if the mountain lion is nearby. Place yourself between the cat and any children. If the child is small, pick her up. If the cat becomes aware of you, it will likely run, but it might not. Stay calm. Not every mountain lion that doesn't immediately flee at the sight of a human being is out to get you. They are intelligent, curious animals and often wait to be sure that you have seen them before making any decisions about what to do. A curious animal may just watch you, ears erect, head high. It may even sit down and peer in your direction. Or it may lie sphinxlike next to some brush if it thinks you haven't seen it.

Not all curious behaviors are signs of imminent danger, but realize that curiosity can turn into hunting behaviors if the cat decides that what it's studying is potential food. Definitely do not allow a mountain lion to act curiously if you have small children with you—this is a line you absolutely should not cross. Generally, all you need to do to dissuade curiosity is to make sure the cat knows you are human. Speak, yell, make noise, and generally the mountain lion's trance will be broken and it will retreat at speed. Some curious animals may approach you, and in such cases only you know how close is too close for comfort. You can generally end the engagement anytime you like by becoming loud and aggressive.

A hunting mountain lion generally drops its head low, slinks or stalks forward, or places itself in a position ready to pounce. Its ears may be laid flat if it's trying to remain invisible, or be forward as it tracks its target. It may twitch its tail as well, especially if it hasn't begun to stalk forward. You should treat these situations as if an attack is imminent. Do not run or act erratically when faced with a curious or aggressive mountain lion. Do not turn your back on it. Do not bend down if it is close enough to reach you in a single bound. The best strategy in these encounters is to become the aggressor rather than a passive victim. Stand tall, wave your hands, yell, and clap your hands. While yelling and clapping, take several swift steps toward the cat—this is a proven method to halt the approach of numerous large cat species.

Make sure it knows you are a human being. Make sure it knows you are no easy target. Make sure it knows you are dangerous. If you have the time and space to reach down and pick up rocks and sticks, arm yourself and throw anything you can at the cat itself—water bottles, rocks, sticks—something that will make it think twice about approaching any closer. Give 'em hell—this is no time to be timid. Back away slowly only if your own threats are ignored, or if the mountain lion is far enough away that it won't interpret the behavior as vulnerability.

If you or someone in your party is attacked by a mountain lion, fight back, and fight hard—hit it, beat it, stab it, poke it in the eye with something sharp (seriously), kick it, wrestle it. Use rocks, sticks, knives, fists, and feet. Keep pounding away until you get a reaction, as cats focal-lock on their prey and often require substantial abuse before they break their concentration. Then they generally run. Fight as long as it takes to end the altercation, and when the cat disengages and walks or runs away, retreat to safety as quickly as you can, without leaving yourself vulnerable to attack from behind. Check behind yourself often just in case the cat is following. If it doesn't close the distance a second time, keep moving but carry rocks and sticks and have them ready.

It is rational to experience fear around large predators that could potentially hurt or kill you, but it is equally irrational to let that fear stop us from doing the things we love. Most people, however, base their fear of mountain lions not on what mountain lions do, but upon what they are capable of doing. Mountain lions are indeed intimidating in form, but their natural behaviors rarely justify our fear of them. In other words, the biological component to our fear of mountain lions is completely rational, but when it is inflated by cultural baggage, it becomes irrational. Take the time to understand the risks associated with mountain lions, to understand the animal, and to make good decisions to reduce the risks as much as possible. We can live with mountain lions with only very infrequent negative interactions. We already do. Most of us will live our entire lives and never even be lucky enough to see one.

CHAPTER 3
Of Lions, Pets, and Livestock

It was July 2018, and Hanako Myers and Marko Colby arrived home to Midori Farm in Quilcene, Washington, relaxed and happy after a short vacation to the coast to unwind after a hectic spring seedling season. They'd left their farm in the hands of trusted staff and another couple experienced in managing farms. Their vacation afterglow abruptly faded upon news that, in their absence, a mountain lion had been raiding the chicken coop uncontested. Their mobile chicken-coop-on-wheels had been parked for the previous week behind the barn on an isolated stretch of the property, out of the way of daily farm operations. They hadn't considered that this arrangement might provide easy access for predators.

With growing anxiety, Marko inquired about their house cat, Mimi. The caretaking couple confessed that they hadn't seen Mimi in several days, and had failed to connect this with the arrival of the mountain lion. It was still early evening, so Marko and Hanako went to look in at the chickens themselves. There, they met the mountain lion, lying along the woodline near the coop, boldly surveying its claim. The couple took to the nearby picnic table to have a better look; the cat acknowledged their presence but remained unperturbed. They agreed immediately that the situation was unacceptable. A mountain lion should not be allowed to take up casual residence and eat their chickens at its leisure. Marko dropped from the table to retrieve his .22 rifle from the house while Hanako further scrutinized the scene. Several panicky, agitated chickens had escaped the three-foot electric fence surrounding the chicken coop and were wandering nearby.

What happened next felt as surreal as it was unexpected. Hanako dropped off the table to secure a wandering chicken at the same instant that a streak of rusty orange-brown flashed across her line of vision. The lion hardly made a sound as it snatched the chicken from her outstretched arms and retreated back to the woodline, twenty yards away. The incident was over before Hanako had time to scream.

Marko appeared in the next moment and fired repeatedly at the ground near the big cat, hoping to scare it off. The mountain lion, however, remained in place. Now Marko could have killed the animal had he chosen to. State laws permit residents to dispatch any mountain lion actively threatening or attacking pets, livestock, or humans. Investigating authorities also often interpret "threatening" behaviors loosely, siding with the shooter in the absence of contradictory evidence. This may seem harsh, and, inevitably, some mountain lions are killed simply for being too close to people and livestock. But killing marauding lions has long been an acceptable practice, and it works for rural residents who prefer to caretake their own properties and belongings.

Midori Farm, however, was an organic farm with a mission that included protecting pollinators and encouraging the presence of wildlife. The loss of a few chickens to predators was a compromise that Hanako and Marko were willing to make in order to share the world with a mountain lion. Nevertheless, they took other steps to protect their farm and discourage the lion's return. First, Marko called their neighbors to alert them to the big cat's presence and to ask their advice. Their response was unanimous and fervent: *You must shoot the cat!*

But Marko did not shoot the mountain lion. Instead, he moved the chicken coop to the other side of the barn and closer to the areas of the farm where people often worked. That simple change was enough to stop the cat from killing chickens. Nevertheless, over the next few days the cat still lingered in the same patch of shrubs and trees. Uncomfortable with the continued presence of the mountain lion, which sunned itself where everyone could see it along the edge of the woods, Marko called in reinforcements armed with fireworks.

Three days after his initial encounter, Marko and his accomplice, Scott Britton, found the lion napping just inside the woodline along a trail carpeted with chicken feathers. They discovered a cache of dead birds, which explained the continued presence of the lion, as well as the remains of the house cat, Mimi. Using an improvised mortar, they lit a firework, dropped it into the tube, and launched it toward the cat with a smoky whiz.

Boom! When the smoke cleared enough to see, the lion was up and moving. They pursued and launched a second firework at the fleeing animal. This one struck true as well, and sent the cat into full flight mode. Amid a burst of sparkling light, they made out the cat's leap, the swishing tail, and then it was gone, never to return to Midori Farm. Unfortunately, though, it visited a neighboring farm to the north several days later.

Mountain lions eat poultry, livestock, and pets as well. They primarily kill goats and sheep when they kill pets or livestock, but also cats, dogs, chickens, domestic turkeys, llamas, miniature horses, and even the occasional cow or horse. In March 2016, P-22, the mountain lion with a Facebook page and a large fan base, broke into the Los Angeles Zoo on the edge of Griffith Park and killed a koala bear. Mountain lions are intelligent carnivores, quick to seize opportunities for an easy meal. In order to transition to strategic coexistence with mountain lions, we must understand their behavior and the options available to us for protecting livestock from lion predation, and we must accept the fact that even despite precautions, mountain lions will kill the occasional pet and ranch animal.

With the growth of mountain lion populations in the late twentieth century, greater livestock and pet losses were inevitable, along with the predictable rise in retaliatory killing by ranchers and farmers. The mental leap from retaliatory killing to preemptive killing as a strategy to reduce any perceived or real risks posed by mountain lions to personal holdings was small. No one responds well to threats to their livelihood, and people have the right to defend what is theirs. On the other hand, it's equally within reason to expect—and even demand—that people with livestock take greater responsibility for their animals, and to take appropriate precautions to protect them.

The Oxford English Dictionary defines *depredation* as "1. The action of making a prey of; plundering, pillaging, ravaging. An act of spoliation and robbery. 2. Destructive operations."[1] State and federal wildlife agents define the term with specific legal implications: When carnivores kill domestic animals or any animal that they wouldn't normally encounter in their customary habitats, they commit *depredation*. And as livestock numbers increased in North America, so did opportunities for mountain lions to kill domestic animals. Today, the United States is home to an estimated 9 billion chickens, 94 million domestic cats,

90 million dogs, 78 million adult cows producing approximately 34 million calves each year, 75 million pigs, nearly 3 million goats, and more than 5 million sheep, as well as uncounted millions of turkeys, horses, llamas, and other domestic species.

Killing mountain lions for depredation is common practice throughout western North America, even in protected populations. In California, which is representative of other Western states, the process is straightforward. When someone finds the fresh remains of their beloved goat and suspects a mountain lion was at fault, they notify state wildlife officials. State wildlife personnel are required by law to assess the evidence within forty-eight hours.[2] If they concur that a mountain lion was the perpetrator, they provide the livestock owner the opportunity to request a ten-day depredation permit. This permit provides the livestock owner or pet owner permission to enlist the aid of trained hounds and to initiate the pursuit of any mountain lion detected within a one-mile radius of the place where the mountain lion killed the owner's goat or house cat. If the hound-hunter locates a fresh trail, or several as might be the case, they can pursue these mountain lions within a circular area with a ten-mile radius around the carcass of the dead animal, assuming owners of any private property they cross grant them permission. Alternatively, someone can come in and bait box traps to try to catch the lion on site; in other states, the use of snares is permitted as well. Once the mountain lion is killed, the pet or livestock owner must report the outcome to the state wildlife agency within twenty-four hours, as well as arrange for the transfer of the dead mountain lion to agency personnel for further investigation.

"A pet owner's nightmare, their dog or cat being eaten by a mountain lion, appears to happen with some frequency, according to a new report from the [California] Department of Fish and Wildlife," wrote sportsman Tom Stienstra in 2016, in his sensationalist *San Francisco Chronicle* article "Study Finds Mountain Lions Are Feasting on House

Pets."[3] Media sometimes fuels the circulation of misinformation about the frequency of mountain lion depredation on pets and livestock. And unfortunately, this misinformation continues to spread.

"If you live in California and are looking for a lost pet, there's a fair chance, unfortunately, that it was an easy meal for a lion," wrote Ben Romans for *Field and Stream*.[4] His source for this commentary? The inaccurate interpretation of a state report presented by Stienstra's *San Francisco Chronicle* article. Stienstra's misrepresentation of the facts rippled out across the United States in copycat articles, maligning mountain lions far and wide.

The report that spurred the original article was entitled "Report to the Fish and Game Commission Regarding Findings of Necropsies on Mountain Lions Taken Under Depredation Permits in 2015." It was a brief annual summary of depredation permits issued by the state, and of the characteristics of the mountain lions killed as a result.

Stienstra summarized the report in these words: "Of those 107 lions [killed], the stomach contents of 83 were analyzed, and 52 percent were found to have eaten cats, dogs, or other domestic animals. . . . Only 5 percent had eaten deer, which are supposed to be their favorite prey, but are harder to catch than house cats." He continued with the figure that "18 percent had contents that were too digested to be identified. If pets also accounted for a good share of that 18 percent, that would mean more than 60 percent of the lions in the study ate cats, dogs and other domestic animals."

His synopsis was inflammatory and inaccurate and meant to feed upon people's fears. First, the actual report was a summary of animals killed by the state under depredation permits, not a study of mountain lions more generally. Depredation permits are only given to people when CDFW confirms that a mountain lion kills their pets and livestock, as discussed above. For this reason, every mountain lion that was

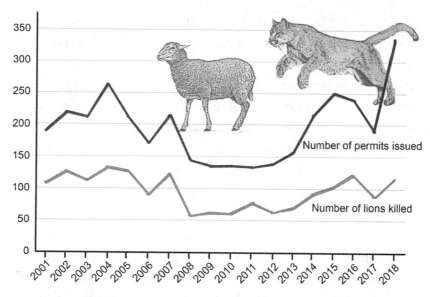

The number of depredation permits issued in California from 2001 to 2018, and the subsequent number of mountain lions killed as a result. Note that the number of permits issued is always higher than that of permits filled. (Data from the California Department of Fish and Wildlife, "Mountain Lion Depredation Statistics Summary," https://wildlife. ca.gov/Conservation/Mammals/Mountain-Lion/Depredation.)

killed should have had domestic animals in their stomachs, not just 52 percent. Stienstra presents that 52 percent figure and then leaps to "more than 60 percent" as an accusation that these 107 mountain lions are representative of the diets of all of California's mountain lions, when in fact they represent the tiny sliver of mountain lions that were accused of and executed for killing domestic animals. Stienstra should have been outraged that 30–48 percent of the big cats killed under depredation permits in California in 2015 may have been innocent—killed for being in the wrong place at the wrong time.

Further, the state report makes no distinction between what kinds of domestic prey were found in the stomachs of mountain lions. Yet

Stienstra assumed they were mostly cats and dogs. In fact, cats and dogs have only accounted for about 8 percent of all California depredation permits issued since 1973.[5] Two-thirds of permits are issued for sheep and goats killed by mountain lions instead. Media can do real harm to mountain lions and other wildlife, because they influence the belief systems that permeate society. Media, however, are not completely to blame for stoking fear of mountain lions. Sometimes, agency reports exaggerate mountain lion depredations as well.

Of the roughly 112 million cows in the United States in 2015, the US Department of Agriculture reported that 3.6 million died from disease and weather, and that 13,384 were killed by mountain lions; therefore, mountain lions kill about 0.01 percent of the total cows in our country each year. Not that it can't be devastating for small farms to lose a single heifer, but overall, mountain lions pose little threat to cattle. Even so, the USDA exaggerates losses to mountain lions, in part because they do not have a system to verify losses reported by ranchers. In their 2015 report, for example, they reported that mountain lions killed 3,217 cows—24 percent of the total number of cows reported as killed by mountain lions—in states *without mountain lions*. In Michigan alone, nonexistent mountain lions killed an astounding 1,471 cows.[6] USDA numbers do not match up with state agency reports in states with mountain lions, either. In 2015, Colorado's state agency biologists confirmed that mountain lions killed 64 livestock—that is *all* kinds of livestock—whereas the USDA reported that mountain lions in Colorado alone killed 208 cows.[7]

By comparison, mountain lions kill more sheep, but the exaggerations in how many sheep they kill are similar. The USDA reported that mountain lions killed over 9,000 sheep in 2014—about 0.14 percent of the total sheep in America.[8] This is still a remarkably small number. Mountain lions were reported to have killed more than a thousand sheep in states *without mountain lions*. In Missouri alone, mountain lions were ludicrously reported to have killed 834 sheep. How the sheep in Michigan escaped unscathed without a single loss to mountain lions remains a mystery, given the rampant cow-killing big cats reported above.

It is irresponsible to spread misinformation about the frequency with which mountain lions kill pets and livestock. It exaggerates not just the actual problem, but also the potential threat of losing pets and livestock to mountain lions. In other words, misinformation fuels fear among pet owners, ranchers, and hobby livestock owners. Fearful people are more likely to act irrationally or to develop grudges against mountain lions even while lacking any direct experience to support their beliefs.

The Midori Farm mountain lion, whose exploits were described at the start of the chapter, appeared on a neighboring property to the north just several days after being chased off by the concussive booms and burning cinders of fireworks. He killed an unprotected goat during the night, and Marko's neighbors decided to move their remaining goats to a safer location. They were in the act of securing one goat in the bed of their pickup truck for transport when the lion reappeared to tackle a second goat tethered to the back of the vehicle. The cat's audacious attack under the mid-morning sun suggested that it was the same animal that had visited Marko and Hanako's farm just days earlier. These livestock owners, however, were less forgiving, and they immediately called another neighbor who appeared to dispatch the cat. With the concussive report of a rifle, the people of Quilcene, Washington, went back to their peaceful lives in the shadow of the rugged Olympic Mountains.

This raises two related assumptions people make about mountain lions that kill domestic animals: (1) a mountain lion that kills livestock will continue to kill livestock; and (2) if you catch and move an offending lion instead of killing it, you just move the problem and it will continue to kill livestock in the new location. There is scant data to back up either of these assumptions.

M68, for example, was a young male mountain lion caught and fitted with a GPS collar north of Jackson, Wyoming, in late 2012 by Panthera's Teton Cougar Project. At the time, he weighed 101 pounds and was covered with porcupine quills, which the team carefully removed before setting him free. They estimated his age to be slightly less than two years old—that magical age during which mountain lions typically search for a territory to call their own, and during which they refine their hunting skills and survival instincts. M68 only survived that winter because he shared the kills made by other mountain lions in the area. But as winter turned to spring, he began killing more animals on his own. He killed coyotes and badgers, beavers and porcupines. He even began killing deer as they drifted back into the system after disappearing from the mountains for the winter season.

In the first days of June, he killed an adult mule deer in the Buffalo Valley east of Moran, a meal that should have fed him many days. But it was promptly stolen by a grizzly bear. He responded not just by abandoning the carcass, but by abandoning the entire mountain range. He traveled 150 miles south in search of new lands in which to establish his territory, perhaps far from grizzly bears that might steal his food. He walked the full length of the Wyoming Range before finding himself without cover, surrounded in natural-gas wells, sagebrush, cattle, and small bands of pronghorns. He hunkered in a rural highway culvert for several days while exploring the area. He killed a porcupine, then two pronghorns, and then—perhaps put off by the sagebrush ocean before him—he looped backwards and retraced his steps back into the heart of the Wyoming Range.

On public lands managed by the US Forest Service, he ran into large, free-roaming sheep herds, grazing legally in a heavily taxpayer-subsidized lease system benefitting the livestock industry. He killed one, and dragged the carcass into cover where he ate in relative peace for several days. Livestock owners can request compensation from the state of Wyoming if they lose livestock to predators, even on public lands. They are paid three and a half times market value of their livestock on the day it died if the predator was a mountain lion or bear, and seven times market value if it was a wolf. They can also request that the predator be killed at the expense of Wyoming's general fund. M68's transgression, however, went unnoticed by the shepherds on site.

Given the distance the research team needed to travel to keep up with M68, they tended to wait several weeks between trips south to see what he'd been up to. By the time they found the remains of the sheep, M68 had already moved on and killed four more prey—two mule deer, a beaver, and a pronghorn. When the team reported the depredation to the Wyoming Game and Fish Department, as required by their research permit, the state agency decided to let M68 go free. The team tracked him for another three months before removing his collar. He did not kill another sheep during this time. This is one example of a mountain lion that killed livestock just once; the Midori Farm provides a potential counterexample in which the lion was a repeat offender. Dr. Mat Alldredge and his team tracked a handful of mountain lions that killed pets or livestock along Colorado's front range. Two were repeat offenders within six months of killing their first domestic animals, and two never killed pets or livestock again. Several others committed one or two more depredations during the four to six years they were monitored.[9]

These examples are just stories—everyone has a yarn or two to share, passed down among ranching communities, growing in each telling. In general, however, it seems that few cats that kill livestock become fabled "livestock killers," and that most cats will only occasionally eat domestic prey if the rare opportunity arises. Hungry mountain lions, meaning those that haven't eaten in some time, take greater risks and are more likely to hunt in suburban environments where they might encounter the abundant pets, raccoons, or even deer that live there.[10] Otherwise, we lack enough examples to determine clear patterns of depredation behavior. One reason for this dearth of data is simply this: we quickly kill mountain lions that kill pets and livestock. State wildlife agencies are not inclined to give an offending mountain lion a second chance in order to study its behavior. In our litigious culture, state agencies fear they would be sued if an offending cat were freed and killed livestock again. So rather than kill mountain lions that attack pets and livestock, might it be better to move them instead?

State wildlife agencies frown on this solution: "Mountain lions shall not be captured and translocated under any circumstances."[11] "Moving problem mountain lions is not an option. It causes deadly territorial conflicts with other mountain lions already there. Or the relocated mountain lion returns."[12] Agencies routinely issue such bold statements, but they are based on a single scientific study, decades-old, comprising a small sample of cats, four of which successfully established new territories and hunted only wild prey.

In 1989 and 1990, researchers working in the White Sands Missile Range in New Mexico moved fourteen mountain lions nearly 300 miles north to the Sangre de Christo Mountains near Colorado.[13] Eight of the fourteen headed south in the direction of their original homes on the missile range. Two males made it all the way home, one in six months, the other taking a full year and a half. Nine of the cats died over the next two years, and five were killed by other mountain lions. One lost his collar, so his fate was unknown. Four cats successfully established new home ranges and integrated into new mountain lion communities. Two mated and produced kittens near their release site in the north during the time period that they were monitored. Based on these facts, states argue that we know everything we need to know about mountain lion translocation.

The burgeoning population of Florida panthers, however, provides counterevidence and suggests that we can successfully move mountain lions. Biologists trial-tested introducing seven Texas mountain lions to northern Florida in 1989. The lions settled into territories and began doing what mountain lions do, killing deer and wandering around. Everything seemed a great success until the start of deer-hunting season and an influx of human activity in the woods.[14] The cats dispersed widely in response, moving into urban centers and areas with cattle operations. So the state wildlife agency captured and removed them before they caused any trouble.

Six years later, after great debate and resistance, biologists tried again. This time, they were more strategic. They captured eight young female mountain lions from west Texas and shipped them to southern Florida, where mountain lions inhabited larger, wilder tracts of land but suffered from inbreeding. The females readily settled into their new homes, produced kittens with resident male mountain lions, and were then recaptured and removed from the population. Their progeny are the mountain lions of the Everglades today.

In other countries, large carnivores that kill livestock—including bears, leopards, and tigers—are routinely relocated. Success rates, however, are soberingly low. Across fifty international studies, carnivore translocations were deemed effective in about 42 percent of cases.[15] Of the translocations that failed, humans were responsible for more than 80 percent of carnivore deaths, primarily by shooting or accidental car collisions. Animals that were moved almost always wandered away from their release sites, often long distances. Some even returned to their capture site, many miles away.

The difficult truth is that moving mountain lions will mostly fail. So we must ask ourselves: Is saving one in three mountain lions enough to justify the costs of catching and transporting them? Right now, transporting mountain lions is costly and politically complicated, and so agencies are reluctant to try. When you add fear of litigation, most agencies balk. Nevertheless, some states are actually trial-testing translocations of mountain lions caught in suburbia or other places where they are unwelcome, as long as they haven't killed any pets or livestock. They are doing so because of changing public perceptions about state agencies, and also external pressure to keep mountain lions alive when they can. With some strategic planning, agencies might be persuaded to move young mountain lions that commit a single depredation, as young animals are more likely to succeed in integrating into a new mountain lion community. As part of the same strategy, older and obviously unhealthy animals could be euthanized immediately.

Given the economic and political burdens of moving offending animals that commit depredation, perhaps it is easier to preemptively protect pets and livestock instead. One long-held argument is that a simple means of protecting livestock is to chase mountain lions with hounds during managed hunting seasons. The theory is that hunted mountain lions learn to fear humans and anything associated with them. Hunting therefore increases pet and livestock safety.

Recent data, however, refutes this logic.

Many Western states increase mountain lion hunting to alleviate livestock conflict, or at least to convey to the public that they are aware of the problem and are trying to alleviate it. Wildlife management is as much, if not more, a social enterprise than a biological one. For example, the Oregon Cougar Management Plan states, "Several factors can trigger a target area [where mountain lion hunting is increased significantly], including the number of cougars being killed for livestock

A portrait of a subadult mountain lion, the sort most likely to have a negative encounter with people and livestock, but also the sort with the best chances of surviving if translocated following an incident.

damage or public safety concerns in the area (indicating a high level of conflict once it reaches a certain threshold) and deer and elk populations falling below management objectives."[16] (The poor justification for killing mountain lions for deer and elk are discussed in chapter 4).

The problems with preemptively culling mountain lions to protect livestock are twofold: (1) there is no evidence that it works, and (2) hunting may increase rather than decrease mountain lion predation on pets and livestock. Following the initial burst of mountain lion research that occurred in the 1970s to facilitate some knowledge of the animal to support mountain lion management, many states concluded that hunting was not helping livestock. As early as 1983, Wain Evans, then assistant director of the New Mexico Department of Game and Fish, recommended reimbursement programs to his state legislature to aid ranchers, rather than hunting. "Efforts to reduce depredations on livestock and wildlife through cougar hunting and control on problem areas has failed," he explained. "Management should take advantage of the cougar's self-limiting potential by allowing development of stable social structures over most of the occupied range . . . [and] . . . reimbursing ranchers for at least part of their loss."[17] Today, ironically, New Mexico has one of the highest mountain lion hunting limits of all Western states, suggesting that Evans's recommendations were lost on New Mexico's politicians.

Hunting may also result in unintentional consequences predicted by "social chaos theory," sometimes called "juvenile delinquent theory." Some researchers argue that where mountain lion hunting is heavy-handed, the older, larger animals are culled from the population, and that youngsters flood in from surrounding country to fill the gaps. Young animals behave most unpredictably and are more likely to run into problems with people on the edges of suburban and urban neighborhoods across North America. The difference between the idea that hunting may increase negative interactions between mountain lions and people, as discussed in chapter 2, and the idea that hunting might increase negative interactions between mountain lions and pets and livestock, is that there is more evidence to support the latter.

Dr. Rob Wielgus is a controversial figure in Western carnivore politics, and nowadays is more often seen wearing leather, straddling a Harley, and carrying a gun. He travels armed because he's experienced several threats to his life. Wielgus was forced to leave his position at

Washington State University, in part because he was the victim of unsavory politics surrounding wolf-killing to protect livestock. Years earlier, however, Wielgus ran a productive research lab for the university, where he was involved in several large, long-term studies of Washington's mountain lions. Over the course of his work with mountain lions, Wielgus became a proponent of juvenile delinquent theory.

Among the many topics they explored, Wielgus and his team sought to determine what might drive mountain lion–livestock conflict in Washington. Not unexpectedly, they discovered that the density of mountain lions in an area was in part linked to the number of conflicts in that area. In fact, they calculated that for every individual mountain lion you add to an area, you increase the likeliness that a complaint will be filed by a pet or livestock owner by 3 percent.[18] Add more lions, and the chances of conflict occurring continue to rise.

More startling was their discovery that the number of mountain lions killed during the previous hunting season was far more important in predicting the number of complaints about mountain lions in an area than was the actual number of mountain lions. For every mountain lion killed the previous season in an area, the likeliness that a complaint will occur in that area *increased* by 36 percent. Killing a mountain lion had more than ten times the impact in determining the likelihood that there will be conflicts in an area as compared to adding one more live mountain lion to that same area.

"For every large resident male [mountain lion] killed, two or three young guys came to the funeral," Wielgus explained. Following juvenile delinquent theory, Wielgus and his team found that "These young cougars were responsible for increased complaints."[19] These results turned everything we thought we knew about mountain lion hunting on its head, and many managers were quick to dismiss it. However, Rich Beausoleil, bear and cougar specialist for the Washington (State) Department of Fish and Wildlife, embraced the findings rather than balked at them.

Beausoleil redistributed mountain lion hunting across Washington State to better avoid creating areas where lots of cats would be killed, resulting in voids that lead to increased problems between young mountain lions and livestock. Beausoleil didn't stop hunting mountain lions in Washington; he and his team improved how they did it.

Researchers in British Columbia conducted their own work and confirmed the findings of Wielgus and his team. Kristine Teichman, Bogdan Cristescu, and Chris Darimont tested what might explain mountain lion attacks on pets and livestock over a thirty-year time span across the entire province of British Columbia.[20] Just like Wielgus, they found that the number of mountain lions killed in any given year or the year previous best explained the number of mountain lion incidents in a particular place. The pattern was especially strong for incidents involving young male mountain lions, providing additional support for Wielgus's funeral analogy. Ironically, mountain lion incidents occurred least often in southwest British Columbia near the city of Vancouver. This is the area where most people reside in British Columbia, providing the highest opportunity for mountain lion–human conflict. It is also the area where mountain lion hunting is least practiced.

And this brings us back to a point made at the start of this section: mountain lion management is more a social enterprise than a biological one. Even if hunting doesn't decrease the number of livestock killed by mountain lions, it builds social capital among the communities suffering losses. People in these areas see their agencies listening to their concerns and trying to help them, even if the strategy is misguided in the end. Therefore, reducing problems between mountain lions and pets and livestock requires an educational component in addition to any actions on the ground, so that people better understand the links between conflict and hunting.

Where wild prey is abundant, most mountain lions will not become repeat livestock offenders. In areas where wild prey is less abundant,

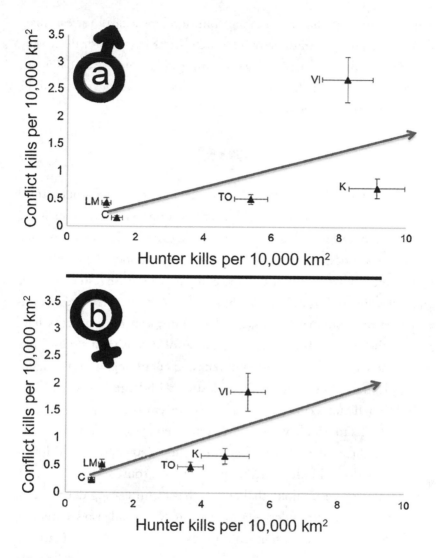

This graphic illustrates the positive relationship between the number of (a) male and (b) female mountain lions killed in an area, and the number of mountain lion incidents with pets and livestock in the same areas, across British Columbia. Modified here with the addition of trend lines. (Originally published in Teichman, Cristescu, and Darimont, "Hunting as a Management Tool?," Fig. 3. Used with permission of B. Cristescu.)

or in the rare case that a mountain lion develops a taste for easy live-stock prey, killing them directly is a much better strategy for decreasing depredations than increasing legal hunting in the area. Removing the perpetrator ends the behavior and, when done by state agencies, can be done swiftly at any time of year.

Here in the United States, we subsidize the killing of predators instead of the protection of livestock. As a result, killing mountain lions is the cheapest and easiest solution for livestock owners. Europe, however, provides a wonderful role model in managing livestock depredation in wolf country. Governments provide compensation to livestock owners who lose animals, but only if the livestock owner has taken precautions to reduce the potential for losses. Many European governments have gone further now, subsidizing the creation of defensive structures like corrals and electric fences, and in purchasing guard dogs and paying for shepherds, to support ranchers in coexisting with large carnivores.

The United States could adopt a similar system and only grant depre-dation permits to people who build livestock enclosures, or utilize other means to reduce the chances of mountain lion depredation. California has already adopted a three-strikes policy in the south, where livestock owners must twice try nonlethal approaches to mitigating depredation before they can request a depredation permit. We could also secure state and federal funds, or even nonprofit funds (as modeled by the National Wildlife Federation in the preface), to assist those in need and to pay for infrastructure to protect livestock from mountain lions. We can and should be part of the solution to the cougar conundrum.

Other options are available to 4-H families and large ranching oper-ations. If one has a small number of livestock—fewer than fifty—the gold-standard strategy to protect them is to corral them at night in an enclosure with a roof. Livestock owners debate how high a fence should be to keep out mountain lions, and it would certainly need to be high— perhaps twelve to fifteen feet. But it's a better strategy to build an enclosure only eight feet tall, but fully covered with chain-link, wood, or other substantial material.

The other proven strategy, as modeled by Marko and Hanako in the story that opens this chapter, is to graze and house livestock and poultry away from areas that provide easy access to mountain lions. Keep livestock as close to places of human activity as possible. Avoid leaving them near woodlines, near riparian zones with shrubs and trees, or along borders with larger wild lands that provide easy travel routes for predators to move undetected.

For medium-sized flocks and herds, or even as a complement to corrals with short walls, foxlights provide great flexibility. Foxlights are solar- or battery-powered strobe lights widely available on Amazon and other local sources; they flash in random patterns and different colors. They can be attached to fences or erected on posts, and they work to keep mountain lions at bay, as well as other critters out of attics—and in lots of other applications besides.[21]

For ranches caretaking hundreds or thousands of livestock, guardian animals and permanent shepherds are the best option. For cattle, people employ a breed called Raramuri Criollo cattle that are highly aggressive and confront intruders that enter stock pastures. These have proven highly effective in protecting cattle in Latin America from jaguars and mountain lions. For sheep, guard dogs are better protection than shepherds, though both provide benefits. Great Pyrenees and Turkish kangals are two dog breeds that have proven effective against mountain lions.

Sheep guard dogs may look like oversized teddy bears, but ranchers report that they can halt sheep losses overnight. Guardian dogs are as aggressive to sheep rustlers and feral dogs as they are to mountain lions. Maintaining livestock guarding dogs requires work, time, and cost

A large white guard dog and his adopted sheep herd. (Photograph courtesy of Beth Wald)

of care, but a single pair of dogs can protect hundreds and perhaps thousands of sheep in contained areas. For larger flocks, more dogs are needed. Some ranchers supplement the Pyrenees with smaller, faster dogs that can detect mountain lions faster and alert the larger Pyrenees where they are needed. In the end, the investment is well worth it. Guardian dogs save money for both ranchers and the state agencies that pay compensation for losses. Simultaneously, they protect mountain lions that might be killed in retaliation, or preemptively to protect the flock.

CHAPTER 4
Sharing Prey with Mountain Lions

F18, an adult female mountain lion followed by the UC Davis Wildlife Health Center in eastern San Diego County, crouched sphinxlike beside a boulder, surveying the hillside below. The slope fell steeply east to a flat valley floor several miles wide, before rising up again in the west toward the town of Julian. The boulder she chose appeared unremarkable. There were many like it scattered across the western slope of the Cigarette Hills just north of the US border with Mexico. This particular rock, however, placed F18 above a travel route utilized by deer for their commute each morning and evening between the cooler cover offered by the riparian area on the eastside of the Cigarette Hills, and the area on the west side where fresh grass burst forth between desert shrubs after recent rains.

As dusk darkened to night, a young mule deer buck made his way down the trail toward the forage below. F18 flattened herself to the ground and pinned her ears back against her skull. She stretched out her neck to accentuate her serpentine length, and stilled her tail. Every muscle was poised and ready.

The deer was alert and quick to see the shadow that unfurled from next to the rock above him. He leapt into a stot, a rapid gait in which

An intimate portrait of a male mountain lion huddled over his kill.

he jumped on all four feet at the same time, as if riding a pogo stick. It's a gait that demonstrated his health and vigor, and it provided him the incredible ability to alter his direction in unpredictable fashion each time he struck the ground. Weaving between vicious barbed cholla cacti, knife-like yucca, and knee- to chest-high boulders, he cut deep, dark gashes in the desert crust in his rush to escape. But the danger had been much too close.

F18 streaked in and angled alongside the fleeing deer. She expertly placed her feet on a small boulder deeply embedded beneath the earth's surface, and launched herself upward onto his back. She gripped her quarry rodeo-style, clinging tightly with sharp, protractible claws. She reached forward in a flash of motion, hooked the deer's head with her claws, and yanked it back so that she might break its neck or use her teeth to crush its throat. Mountain lions have four cylindrical teeth at the corners of their mouths with which they often crush and sever their prey's windpipe, killing by suffocation.

The dead deer crashed down with such force that he broke an antler midway along its length during his collision with a large boulder. F18, remarkably uninjured during the fall, paused over her prey to rest. Then she dragged the deer around to the other side of the rock, where she opened the carcass. She was quickly joined by her eighteen-month-old kitten, F43, but F18 did not allow her to feed immediately. She seized great chunks of deer hair between her front teeth and yanked them out, spitting them to either side. With the skin thus exposed, she then chewed a neat hole at the point where the hind leg met the abdomen. Initially, she fed on the inside of the hind leg. Then she opened the abdomen completely and removed the stomach with surgical precision. She used her carnassial teeth to shear the ribs as cleanly as any pair of scissors cuts paper. Once the chest cavity was exposed, F18 and F43 took turns gorging on the blood-rich organs, rich in minerals and vitamins.

Perhaps no other mountain lion behavior so captures the human imagination as their ability to subdue and kill prey many times their size. And it's no doubt that their unique weapons and hunting prowess make them scary and controversial. Mountain lions are stalk-and-ambush hunters, waiting for or slinking to within striking distance of their intended targets. They are patient experts of camouflage, using landscape features and vegetation as short as eight inches to allow them to flow toward their prey. Their hind legs are disproportionately long and provide them the leaping and running power needed to catch fleet prey. Their thick, muscular tails act as counterbalances as they run, and their floating clavicle bones allow for tight turns while pursuing terrified prey. If they can get close enough to their quarry undetected, they explode into action—and their chances of catching it are excellent. In one study, cougars were successful in thirty-seven of forty-five attempted hunts.[1]

"There's a lot of complications with how we manage mountain lions," began Sy Gilliland, a representative of the Wyoming Outfitters and Guides Association, and owner of one of the largest outfitters in the state. It was July 8, 2016, and the seven members of the Wyoming Game Commission were meeting in Pinedale to review proposed changes to hunting regulations, including those governing mountain lion hunting. "To hear the houndsmen talk about it, they want lower quotas, they want more mountain lions. [dramatic pause] Well, mountain lions eat meat. They eat mule deer at an incredible rate. That's what they live off of." Gilliland leaned in and gripped the pulpit with the intensity of a Baptist preacher. Then he began to weave melodramatic anecdotes to ensnare his audience.

"So I hunt mule deer. One of the ranches I'm most familiar with is in mountain lion hunt area 22. Prior to the big winter die-off we had in 1994, my organization used to hunt out of three different camps on that ranch. We'd hunt ninety mule deer a year. Now we're hunting twelve mule deer a year. We've got an incredible lion problem." (Note that as stated above, the impact of the "big winter die-off" on the change is unclear.) Gilliland deftly transitioned into a story about a mountain lion scaring horses in camp, causing one to impale itself on a fence post. He well knew the power and sway of a tragic tale of a horse dying unnecessarily in Wyoming. He continued with another story about a lion following one of his guides throughout the day as he packed out an elk carcass following a successful hunt, implying that the guide, or perhaps his client's elk, was in real danger. Gilliland was relentless.

"Two years ago," Gilliland continued, "the ranch came off the mountain fifty-five calves short. They probably lost twelve to natural mortality, the rest were from lions [how he could know this is unclear]. . . . So when we start talking about this, and all we're hearing is from the houndsmen, we want more lions, we want more lions, we want more lions—that's at the expense of another whole large sector of sportsman, which is mule deer hunters."

Although they are a small proportion of Americans, the hunting community is an impassioned and politically well-connected group of individuals invested in the conservation of natural resources and, more important, their way of life. As mountain lions have reclaimed former range across western North America and increased in number, sportsmen's angst over lost opportunity and having to share the bounty with recovering large-predator populations followed in perfect parallel. Ultimately, hunters and the wildlife agencies that serve them are concerned about competition for shared resources—meaning competition for the deer and elk and other game species they love to hunt. Again and again over the years, sportsmen's lobbies have put pressure on state agencies responsible for mountain lion management, and these agencies in turn have capitulated to the interests of their constituents and have increased

mountain lion hunting. This makes some sense, as hunters and sportsmen's associations disproportionately pay for state wildlife agency programs (discussed further in chapter 8), and because state agencies in part
operate like businesses that prioritize their paying customers.[2]

One of the main reasons we kill so many mountain lions across
North America is to protect deer, elk, and the interests of those invested
in hunting deer and elk. Thus, resolving the interests of hunters and
nonhunters, and disentangling fact from fiction, are important components of the cougar conundrum. Rest assured, mountain lions are not
eating all the deer and elk, nor are they targeting prized bull elk or mule
deer bucks.[3] Predators do not wipe out the prey species upon which they
depend. For the most part, mountain lions do not cause catastrophic
declines in deer, elk, and other ungulate (hoofed animal) numbers,
either—though, on occasion, they do. More importantly, killing mountain lions rarely increases ungulate populations—though, under specific
circumstances, it can.

Mountain lions are born with hunting instincts, but they aren't born
ready-made killing machines. Their hunting prowess comes slowly and
is honed over years of practice, even long after they have left their mothers. It begins with play when they are barely able to walk. The lives
of kittens are one long series of stalks, chases, and pounces that prepare them to tackle prey rather than siblings and their mother's tail.
Their mother provides them learning opportunities as they grow as well,
wounding prey and then allowing the kittens to dispatch them.

Young mountain lions generally leave their mother when they are
between fourteen and twenty months old, with the odd animal leaving
earlier or as late as twenty-four months of age. They may wander for
a month or a year without a territory and the intimate knowledge of

An adult female mountain lion caching her bull elk in northwest Wyoming. Mountain lions preferentially hunt deer and elk younger than one year old, but on rare occasions like this one they kill adult male ungulates as well.

place that comes with it. Young mountain lions don't know where to look for prey and are prone to try to catch smaller animals and whatever they encounter.[4] Subadult lions also lack the requisite experience, even if they do have the weaponry, to efficiently and safely kill the largest, most dangerous prey on the landscape. Unlucky and unpracticed mountain lions can be maimed or killed in their attempts to hunt adult deer and elk armed with strong legs, sharp hooves, and sometimes antlers.

In general, adult mountain lions predominantly hunt the most common ungulate wherever they live. This might be mule deer, white-tailed deer, elk, the llama-like guanaco in Patagonia, or even domestic sheep in parts of South America. Mountain lions eat just about anything up to the size of an elk, but they also occasionally hunt horses, adult cattle, and moose. F18, the cat described above, hunted mostly mule deer in her high-desert habitat, but also supplemented her diet with diverse small prey including wild turkeys, raccoons, coyotes, gray foxes, and skunks.

Not all mountain lions eat the same prey, and many develop a unique diet favoring a particular prey species. Some eat lots of pronghorns, for example, or porcupines, or bighorn sheep. How a mountain lion develops a distinctive taste for certain prey is unknown, though a few researchers have speculated on the subject. Most believe it begins with a chance encounter between cat and prey, and the cat's snap decision to try to hunt it. If the mountain lion succeeds in killing it, it receives a reward for its efforts—food. Thus, they may take up the chase again the next time they encounter the same sort of animal, because they remember the reward. As a consequence, a feedback loop develops, providing increasing rewards for hunting a particular kind of prey, accompanied by an increasingly specialized skill set that allows them to successfully hunt that prey. To cite one example, Blake Lowrey and colleagues followed an adult male mountain lion in Colorado with a particular taste for beavers. More than any other lion they tracked, this male dropped into valley bottoms and slowly walked miles and miles of stream banks each week in search of chubby beavers with meat rich in fats and minerals.[5]

Mountain lions kill a lot of animals—about three times as many as they need to survive.[6] State wildlife personnel have long been preoccupied with determining how many deer and elk each mountain lion kills in a calendar year, because of the implications for determining the impact of mountain lions on ungulates we value, like deer, elk, and bighorn sheep. Early creative researchers were so obsessed with this idea that they wrestled captive mountain lions into masks trailing long tubes attached to nearby machines, and then set them walking and running on treadmills. They measured each animal's oxygen consumption in order to determine the energetic costs of exercise. Once they knew how much energy a cat needed to get around, they scaled up to how many deer mountain lions needed to sustain themselves each year. Researchers refined these methods several times over the years, but they always underestimated what big cats actually killed in the wild. This is because mountain lions kill more than they need to meet their energetic requirements. They kill enough to compensate for their kills that are stolen by bears and wolves, spoiled by warm temperatures and insect activity, and sometimes frozen into ice, making them inaccessible. Mountain lion kill rates and the number of prey they kill per year are interesting, but these data can become a distraction from the crucial question on everyone's minds: Are mountain lions eating all the deer?

North Americans are fiercely defensive of their ungulate populations and the right to hunt them. As a result, the various state agencies in the United States has invested billions of dollars in research to understand what determines the number of deer and elk on the landscape and how many can be sustainably hunted each year. Countless studies point to weather, which determines the abundance and health of grasses, forbs, and woody plants. Forage quality and quantity, in turn, determine the number of ungulates a particular landscape can support. Weather events such as droughts, which reduce forage for deer and elk, and long, harsh winters, which cause mass starvations, can also substantially impact deer and elk numbers. In short, weather, not predation, is the chief influence on wild ungulates' populations. But other factors play a role as well, such as carrying capacity.

Wildlife biologists and managers often talk about the "carrying capacity" of an area, which simply means the number of animals an area can comfortably support. The carrying capacity is usually limited by a particular resource, such as food. In the case of deer, carrying capacity is the maximum number of deer that can be supported by an area, given reliable forage, good weather, and suitable shelter to protect fawns and adults. When deer populations are at or near carrying capacity, there are *too many* deer, and the deer themselves control their numbers. The animals tend to be in poorer condition because of competition among themselves over dwindling food. Females produce smaller, weaker fawns that die more easily, and they are more susceptible to numerous forms of mortality, such as a harsh winter or a disease outbreak. No one wants deer populations to be at carrying capacity, not even state wildlife agencies or hunters, as no one wants to hunt starving animals with their ribs showing.

When a deer population is below carrying capacity, there is plenty of grass and woody browse to go around. This means the deer are healthier and they produce heftier fawns more likely to survive to become breeding adults. A deer population below carrying capacity supports happy hunters and state agency biologists, as long as the population doesn't drop too low. State agencies and the hunting constituents that fund them seek that perfect balance where deer are numerous, but not so numerous they begin to starve and turn ugly.

When predators like mountain lions are figured into the mix, carrying capacity is critical. If a deer population is at or near carrying capacity, predators do not affect the population, since any deer a predator kills would almost certainly have died anyway from some other cause, like starvation or disease. Biologists call this *compensatory* killing—meaning that predators cause deaths that compensate, or counterbalance, for deaths that would have happened anyway. For these reasons, mountain lions do not impact deer or elk populations that are near carrying capacity. But when a deer population is below carrying capacity, predator kills become *additive*, meaning that they are deaths that were unlikely to have occurred for some other reason if not for the predator that killed it.

Mountain lion predation on deer and other ungulates is often additive, as states' wildlife management tries to keep deer populations below carrying capacity. Additive mortality, however, represents a spectrum of

A mountain lion spots prey and watches intently before descending to begin her hunt.

effects ranging from inconsequential to devastating, depending a great deal upon how close to carrying capacity the population is, and the current effects of weather and forage quality on said population. Additive mortality doesn't mean the deer population is doomed, nor that the deer population cannot increase even under predation.

In general, hunting mountain lions does not help deer or elk herds for two reasons: (1) deer and elk are rarely far enough below carrying capacity to allow predation to have a large effect; and (2) the effects of human hunting on deer and elk overshadow the effects of mountain lions. For example, in a review of female elk survival across forty-five study populations, led by Jeremiah Brodie in collaboration with many other respected scientists, predation was additive but reduced female elk survival by less than 2 percent.[7] When they included the effects of human hunting on these same elk populations, the effects of native predators became compensatory, because they were deaths that would have almost certainly occurred at the hands of human hunters instead. In another meta-analysis of 101 management units in the West, forage productivity was predictably the most important driver of elk recruitment, which refers to the proportion of youngsters that survive to become breeding adults;[8] complex predator systems with wolves and grizzly bears, however, also had a measurable effect on recruitment. In a review of mule deer dynamics conducted by Tavis Forrester and Heiko Wittmer, the authors concluded that predation by mountain lions and other predators did not appear to cause mule deer declines, and for the most part was compensatory.[9] All in all, human hunters kill exponentially more deer and elk than mountain lions, and therefore, as Brodie and his team made clear in their review, the best strategy to aid ungulates is to reduce human hunting—not predators.

Ecologically, the more important questions we should ask are How small does a deer population have to be before mountain lions really impact it? And, how small does a deer population need to be before removing mountain lions might help that population recover? Deer are actually a poor example, because generally mountain lions do not impact deer populations enough to merit management intervention. We've known this for decades—it was among the key findings of Maurice Hornocker's landmark work on mountain lions conducted in the 1960s and '70s. And we've relearned this lesson several times since as well. For example, researchers killed mountain lions and coyotes from 1997 to 2003 in Idaho to see if they could increase local deer numbers.[10] Killing cats allowed more fawns to survive longer, and it did increase the number of does with fawns in the population, but in the end, these differences did not translate into deer population growth. And it's the end result that counts: killing predators did not grow the deer population.

Be careful consumers of agency statistics and how agencies interpret results, especially given that current politics and money play a role in promoting the narrative that predators are causing deer and elk declines.

Research that highlights how many elk and deer mountain lions kill, or emphasizes that mountain lion predation is additive, is meaningless in the absence of data showing the true effect of predation on deer and elk population growth rates. As highlighted earlier, weather and forage quality are the primary drivers of ungulate population numbers, and it is too easy to make mountain lions scapegoats for other difficult issues, such as drought, new subdivisions and natural gas development, that are the real causes of deer population declines across the West.

Bighorn sheep are a better example of an ungulate that can occur well below carrying capacity, so that controlling mountain lions may be warranted. Killing mountain lions to aid bighorn sheep, however, is not without controversy. Sportsmen highly prize the massive, spiraling horns of rams, and when their populations are in jeopardy, state agencies are quick to blame mountain lions. In fact, disease, droughts, impoverished habitats, and overhunting are all factors contributing to the decline of bighorn sheep populations in the West, far outweighing mountain lion predation. Pneumonia, for example, is an unexpected culprit, which wild sheep contract when they rub against domestic sheep that graze on public lands. The key point is that whereas healthy populations of deer and sheep can largely shrug off the predatory effects of mountain lions, populations compromised by drought or disease have fewer defenses.

In 1999, the New Mexico State Game Commission approved a management plan to cull mountain lions in order to aid dwindling desert bighorn sheep populations. Eric Rominger was the biologist charged with overseeing the recovery of bighorn sheep in the state, and also the mountain lion removal program.

Over the initial eight years of the program, ninety-eight mountain lions were removed from three mountain ranges (Peloncillo, Hatchet, and Sierra Ladron). Rominger paid contract houndsmen and trappers $293,450 to kill eighty-three lions, sport hunters killed twelve for free, one mountain lion was killed for livestock depredation, and two more died on roads.[11] Rominger quickly expanded the culling of mountain lions to every area where they were about to release bighorn sheep. Thus, houndsmen and trappers cleared areas of as many lions as possible before Rominger and his team dropped off captive-bred animals, or bighorn sheep caught in the largest herds and transported via helicopter. Rominger believed this was the logical strategy for giving small groups of sheep the best chance of survival.

Perhaps because of public backlash over killing mountain lions, Rominger became bolder about his methods, and more outspoken. The data he shared at conferences and public forums was tantalizing—bighorn sheep in his original study areas lived longer, more productive lives after intensive mountain lion removal was initiated. In a talk he called "Culling Mountain Lions—Some Lives Are More Sacred than Others" at the Seventy-Second North American Wildlife and Natural Resources Conference in 2007, he lamented "the national, and even international, outcry that results from a state wildlife-management agency's decision to conduct lethal removal of top carnivores, such as wolves or mountain lions," as well as public fallout over mountain lion control more broadly. But he remained steadfast in his unwavering belief that it was, at times, necessary.[12] Rominger was a bighorn sheep advocate, and stubbornly set on increasing bighorn sheep numbers to where they could be harvested by sportsmen again in the state of New Mexico.

Many argue that managers are justified in killing mountain lions to save bighorn sheep because we are in part responsible for increased lion–sheep interactions in recent decades. Historically, few mountain lions inhabited desert ecosystems, and thus the dilemma we face today is the result of shifting landscapes and wildlife populations following the dramatic changes wrought by European settlement. By the turn of the twentieth century, cattle and domestic sheep were well established across the United States, and their churning hooves and insatiable appetites had fast changed the face of the American West. Grazing livestock, for example, changed vast desert grasslands into shrublands and chaparral. Indigenous caretakers that had periodically burned the grassland ecosystems to keep them open had been corralled by settlers, exterminated, or been integrated into mainstream society.

Once, these habitats excluded deer and supported pronghorn and bighorn sheep populations that ate grasses and forbs, but over time they became increasingly better habitat for browsing deer that could survive on shrubs instead. As deer populations recovered following the establishment of game laws and hunting seasons in the first half of the twentieth century, the deer had many more places to live than in previous centuries. The new landscape molded by European settlement suited their needs just fine. Mule deer moved into new shrubby desert habitats, including the Great Basin of Nevada and the dry canyons and mountains of New Mexico and Arizona. And with recovering deer there followed the recovering mountain lion populations that ate them, saved by the cessation of bounty programs in the mid-twentieth century. Suddenly desert bighorn sheep were being forced to compete with deer in shared habitat, as well as to share the deer's predators that had followed them into desert ecosystems.

Deer, not mountain lions, are the root of the current bighorn sheep dilemma. Yes, the Europeans brought the livestock that reduced habitat for bighorn sheep, and yes, the settlers introduced diseases that decimated bighorn populations. But today, mountain lions kill bighorn sheep most often where they overlap with deer, the preferred pray of mountain lions. For example, in both the Sierra Nevada mountains of California and the Gros Ventre mountains of Wyoming, mountain lions primarily kill bighorn sheep in winter, when the sheep descend in elevation onto winter range they share with deer and elk.[13]

2016-01-25 12:48

A female mountain lion lounges against a bighorn sheep ram killed on winter range where bighorn sheep overlap with elk, and elk predators, in northwest Wyoming. (Photograph courtesy of Michele Peziol / Panthera.)

Historically, large mountain lion populations couldn't establish and maintain themselves on the dispersed populations of bighorn sheep in desert environments—wild sheep were just too few and too spread out. Mountain lions today, however, sustain themselves on a diet of more abundant deer, and then encounter and kill bighorn sheep in areas where deer have established. And if a band of desert sheep numbers only twenty animals, even the loss of a few is a significant setback.

As early as 1991, several conservation biologists boldly asked whether mule deer and mountain lions should both be removed from desert ecosystems where they were previously absent to rebuild "pristine" pre-European native ecosystems.[14] No one acted on the notion at the time. It was sacrilegious, if not un-American, to talk about removing or reducing mule deer herds. But lions were another matter, and state agencies charged with managing bighorn sheep began to consider the possibilities of killing mountain lions to aid sheep populations.

The science supporting killing mountain lions to aid bighorn sheep is far from bulletproof, however, because the power of statistics only bears fruit when a lot of data is available. Desert bighorn sheep herds are small, and mountain lions only kill a few of them, so statistical analyses provide weak support for the idea that killing cats aids sheep recovery. Rominger also argued that other experimental approaches, in which he might have compared herd survival in groups where mountain lion predation was allowed to go unchecked versus those in which mountain lions were removed, were risky and unrealistic: "I had no inclination to leave half those herds as controls only to find out that, guess what, they went extinct because of lion predation."[15]

On the one hand, Maurice Hornocker, Ken Logan, and Linda Sweanor found that mountain lions killed fewer bighorn sheep on the White Sands Missile Range in New Mexico when managers stopped killing mountain lions to defend sheep. They believed that this was the case because when they stopped killing lions, the mountain lion population recovered and stabilized. Hornocker believed that hunting lions (as discussed in chapters 2 and 3) increased the number of young transient mountain lions in the population, and that young cats were more likely to eat sheep they encountered.[16] On the other hand, when we amass all the small studies of mountain lion predation on bighorn sheep conducted by Rominger and others, it seems clear that mountain lions can hinder the recovery of small sheep populations and, in extreme cases, may even be a threat to their long-term persistence. This very much fits in line with current science as well, which predicts that predators become a threat to prey populations when they fall well below carrying capacity.

Whatever his methods, Rominger succeeded in bringing New Mexico bighorns back from near extinction. In 2011, bighorn sheep were removed from the state's endangered list, for which Rominger is largely credited and was selected for the Wild Sheep Foundation's Wild Sheep Biologist's Wall of Fame.[17] Since that time, bighorn sheep hunting has resumed in New Mexico, and hunters now kill an estimated sixty sheep annually. Culling lions may have very well been part of rebuilding New Mexico's sheep populations, but Rominger's methods were not really designed to test whether it definitively was.

To be clear: killing mountain lions is at best a temporary reprieve for bighorn sheep or any other ailing ungulate species. The better approach to bighorn sheep recovery is to address mountain lions and deer simultaneously; unless we remove the deer, any lion that managers remove will be replaced by another that steps in to fill the void. Ideally, we would restore habitat simultaneously with mountain lion and deer removals in desert systems.

Our goal must be to support an increase in the number of sheep in the population to the point where they can sustain themselves, *even while living with predators*. Because so many species are in need of conservation interventions today, breeding programs meant to bolster or re-establish rare and endangered animals are a booming business around the world. The traditional approach with bighorn sheep and many other species has been to capture some remaining wild animals, breed them carefully in a natural enclosure with minimal human contact, and then release their offspring in small numbers back into the wild. It's a slow process, and it sometimes fails when too many of the young die either in captivity or after they are released.

Many frustrated conservationists are abandoning the old approach of working in semi-natural conditions with carefully isolated animals, and are instead taking up plastic gloves, lab coats, and test tubes to work under artificial lights in warehouses. They are adopting livestock husbandry practices that maximize output rather than quality—a certain percentage of young produced are cycled back to become breeders and the others are readied for release. Such programs have found that they can exponentially increase their output of animals for reintroduction programs, and that in the end, a greater number of naïve youngsters have a better chance of establishing a new population than a smaller number of carefully controlled, more ecologically prepared youngsters produced as before. This may be one solution for the bighorn sheep dilemma that would better allow mountain lions and wild sheep to coexist.

If we believe that killing mountain lions and other predators is at times necessary to save bighorn sheep, woodland caribou, and other rare ungulate species, we need to ask one last question—what is the best way to do it? Should we be killing as many mountain lions as we can, as practiced in New Mexico, or selectively removing only those cats that are known bighorn sheep killers, as role-modeled by California instead?

We know that not all mountain lions hunt bighorn sheep. In fact, bighorn sheep "specialists" occur sporadically and unpredictably among mountain lion populations.[18] So it would seem that targeted removal of mountain lions that hunt bighorn sheep makes more sense, both in terms of its effectiveness in aiding sheep, and ethically, to avoid unnecessarily killing mountain lions. The dilemma, however, often boils down to money, and the reality faced by wildlife management agencies in balancing low cost versus best practices. It is expensive to follow mountain lions near bighorn sheep in order to know their diet, especially when compared to the free services provided by mountain lion hunters should an agency raise mountain lion hunting quotas where sheep exist, or even the costs of focused, contract culling in which the lion population is significantly knocked back.

After eating her deer on the Cigarette Hills, F18 (the mountain lion discussed at the start of the chapter) moved far to the north beyond the typical extent of her territory. She and her kitten parted ways near Palm Springs, and F43 began the perilous journey to search for her own territory. F18 immediately traveled back to her own corner of the world, cutting a straight line southwest across the northwest canyons of

Anza-Borrego State Park. This meant that she dropped out of the mountains north of Anza-Borrego to traverse low-elevation desert canyons, and then again climbed up into the mountains to the west of the state park on the way to the Cigarette Hills in the core of her territory. She didn't make the long trip in one continuous walk, however. She paused for three days high up in one of the state park's desert canyons, inciting serious concern among the researchers who followed her.

Anza-Borrego and the surrounding mountains are a particularly unique ecosystem, with an invisible line floating at about 3,000 feet above sea level. Above this elevation in and adjacent to Anza-Borrego State Park, mule deer live among juniper bushes, oak forests, and pine trees in the cooler ecosystems at higher elevations. Below 3,000 feet lay a rugged, rocky desert landscape, carved out by water and time to create myriad canyons holding surprising oases lined with real palm trees, called California fan palms, and a small, fragile population of desert bighorn sheep. Bighorn sheep are of such concern in California that their numbers are monitored continuously, and mountain lions that partake of them too often are killed by state officials. F18's pit stop fell well below the 3,000-foot threshold, and so everyone assumed that she had killed her first bighorn sheep.

When researchers arrived on site to determine F18's fate, clouds dotted the sky. The moon-like landscape was mostly made of fragments of pitted red rock, interspersed with sparse but hardy vegetation. Tall, skeletal ocotillo plants grew in clumps, intermingled with spiky, waist-height cholla. The area was completely quiet. There were no birdcalls or scratching sounds from lizards' claws on stone. It was as if the desert itself was holding its breath in anticipation of F18's judgment. And then they found it, tucked under a large boulder to hide the carcass from scavengers. F18 had encountered and killed a stray mule deer buck that had wandered down from the high country during the cooler winter weather.

California has led the way in practicing the targeted removal of mountain lions that hunt bighorn sheep, rather than large-scale culling, as practiced in New Mexico. Because mountain lions hold territories based on prey availability, there can only be so many mountain lions supported in any given area—this was one of the first lessons taught to us by pioneering researcher Maurice Hornocker and his colleagues. Thus an animal like F18, which preferred deer over sheep, was holding a place on the landscape where a "sheep specialist" lion could live instead—and so she was the ideal lion to keep in the local population, because she overlapped with bighorn sheep but didn't eat them. Should F18 have died for any reason, the next lion to fill the gap where she lived just might be a bighorn sheep killer. Therefore it benefited wild sheep to keep her alive.

To a large extent, the hunting framework within which we manage mountain lions throughout the West, including culling conducted by Wildlife Services and other contract trappers and houndsmen, is meant to support sustainable deer, elk, and bighorn sheep hunting for the American people. In part, we hunt mountain lions because they are competition for shared natural resources, and we hunt them hard because sportsmen's associations and their constituents pressure state agencies to ensure that we do so.[19] Evidence suggests that our perpetual concern over sharing elk and deer with mountain lions, however, is almost always unwarranted. So is the high numbers of mountain lions we kill to protect our own interests.

Mountain lions, however, are nothing if not resilient. Their numbers swelled when they were given partial protection by state agencies during the second half of the twentieth century. But the trend in recent years is to increase mountain lion hunting as the solution to every aspect of the cougar conundrum. Let us pause then in our exploration of the cougar conundrum to assess how mountain lions are faring in a modern world.

CHAPTER 5
The Great Hunting Debate

In 2017, the Humane Society of the United States (HSUS) released a report titled *State of the Mountain Lion: A Call to End Trophy Hunting of America's Lion.* In their report, they state that mountain lions "are predominantly threatened by human-caused activity, primarily trophy hunting and habitat loss." The report emphasized "rampant trophy hunting" as the greatest threat by far to mountain lions in the United States and Canada, a perspective shared by a large contingent of people who think US mountain lions are in trouble and in immediate need of greater protection.

Concerned that the HSUS report might undermine current mountain lion management, state wildlife agency personnel, representing a different contingent of people who think that US mountain lions are doing just fine, requested that US Geological Society (USGS) biologists conduct a formal review of the HSUS report. In their counter-report, USGS authors concluded that the *State of the Mountain Lion* "fails to serve as a scientifically defensible foundation for management recommendations range-wide or at the State level."[1] State wildlife managers secured exactly what they wanted, which was a resource to refute any claim made by HSUS about mountain lion hunting or mountain lion populations more broadly, and a means of silencing those opposed to current mountain lion management. This is just one of many examples of tit-for-tat exchanges between wildlife advocates and state wildlife managers, but it highlights concerns over opposing opinions about the current status of mountain lions in the United States and Canada. In this chapter we'll pause from our exploration of the cougar conundrum as it impacts us and instead focus on determining how mountain lions are faring under increasing human pressures. People want to know, even though they tend to go about asking indirectly. They instead focus on hunting.

Mountain lion advocates and pro–mountain lion hunting advocates continue to debate whether legal hunting threatens mountain lions in North America. The fight—and it is one—is ugly, and devoid of politeness or tolerance for political differences. On social media particularly, it quickly devolves into mudslinging because it represents fundamentally opposed worldviews. The question gets to the difficult ethical, moral, physical, and spiritual implications of hunting. Animal advocates wonder how one can be human and kill sentient animals. Hunters argue that one cannot be human without participating in the circle of life—which includes taking life to survive.

Ethics and morals aside, the dilemma over hunting mountain lions is important. We need to know whether managed hunting as conducted throughout the West is a threat to mountain lions. First, however, we need to address an issue of semantics. *Threat* can mean many things: chain-smoking cigarettes threatens your health, ice storms threaten telephone wires, and running a preschool of twenty students by oneself would certainly threaten one's sanity. In this context, hunting is a threat to mountain lions, just as a back-alley thug is a threat to everyday people who find themselves in the wrong place at the wrong time. However, when biologists use *threat* in reference to an animal species, they mean something different, something very particular.

The International Union for Conservation of Nature (IUCN) assesses the status of species worldwide, ranking them in six main categories, here ranked from lowest conservation concern to highest: Least Concern, Conservation Dependent, Near Threatened, Vulnerable, Endangered, and Critically Endangered. The IUCN explains: "Direct threats are the proximate human activities or processes that have impacted, are impacting, or may impact the *status* of the taxon [meaning a specific animal]." Mountain lions are currently listed as Least Concern, though their populations in Latin America are declining, and so the question becomes this: Is managed hunting in the United States impacting mountain lions so dramatically that it may exterminate them from portions of their current range, requiring a change in status from Least Concern to Near Threatened or Vulnerable? Following this narrower, biological definition, the answer is no, absolutely not. Hunting as managed today is not a threat to mountain lions in the western United States. One hundred years ago, the bounty system was a threat to mountain lions as it eliminated them from most of their range in North America. This is not the case today.

But this is not to say that hunting doesn't impact mountain lions, or that we aren't overhunting them. Even mortality due to "light" hunting pressure is additive,[2] which (if you recall from chapter 4) means that legal hunting is killing mountain lions that would unlikely have died any other way if not for the hunter who killed them. As we've discussed in previous chapters, too many people think hunting is the solution to every aspect of the cougar conundrum. Even so, the effects of sport hunting and other human impacts on mountain lion populations are still not fully understood. This scientific uncertainty creates questions, concerns, and debate over how mountain lions are faring in modern America.

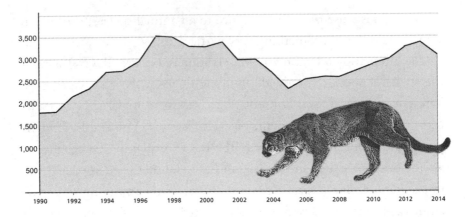

The annual number of mountain lions killed in the US as part of legal hunting from 1990 to 2014. It is interesting to contemplate the potential relationship between this pattern and that presented for attacks on people, found in the figure on page 32.

Annually, we kill somewhere between 3,000 and 4,000 mountain lions in the United States and Canada, mostly through sport hunting, but also for human and livestock protection. Humans are the leading cause of death for mountain lions everywhere in the United States. This is true even in states where hunting is illegal, such as California and Florida, because we also kill mountain lions via vehicle strikes, depredation permits, and other non-sport-hunting causes. What is clear is that hunting, which is practiced across most of the western United States and Canada, is almost always the greatest impact on mountain lion survival and their population dynamics.

It was July 8, 2016, and the seven members of the Wyoming Game Commission were meeting in Pinedale, Wyoming, to review regulation changes proposed by Wyoming Game and Fish Department (WGFD) staff. Commissioners are appointed by the governor for six-year terms—one of the clearest examples of how wildlife management functions more as a political machine than an ecological one. The room was unadorned and felt small for the event, even cramped; agency staff stood in the back of the room near the door and the self-serve coffee stand. The twenty-five chairs set for the audience were mostly empty.

Justin Clapp, large-carnivore biologist for Game and Fish, stepped up to the lectern to present proposed changes to Chapter 42, which outlined mountain lion hunting rules and seasons for the state of Wyoming. Clapp had shaved his beard and suffered a haircut. He gave his presentation in a moderated, professional tone. He used agency lingo like "hunter effort" and "harvest," terminology meant to obscure the fact that he was providing summary statistics about animals killed by hunters in the field. His tone was mesmerizing, and his calm, unruffled demeanor lulled his listeners. But for those who followed his presentation closely, his report about the state of mountain lions in Wyoming was shocking. Clapp provided evidence that the massive, sparsely populated state of Wyoming had succeeded in reducing its mountain lion population through aggressive management. Clapp concluded his overview by emphasizing the "overwhelming success" of mountain lion reduction before detailing the agency's management recommendations for amending Chapter 42, which included reducing hunting pressure in several areas for the first time in a decade.

Heavy hunting can temporarily reduce a mountain lion population in a small area, such as a mountain range. But to significantly reduce the number of mountain lions in an area the size of Wyoming is difficult, requiring heavy, long-term hunting pressure. This is because individual mountain lions live in communities within populations within larger meta-populations in a vast mountain lion superorganism connecting every living mountain lion from Canada to southern Chile. The connections between far-flung populations are maintained by younger mountain lions, which travel astonishing distances to establish new territories, bringing with them new blood and genetic stock to infuse the population they join. Young male mountain lions almost always disperse away from their natal grounds, but females as well may seek a territory in some distant land. This is part of what makes mountain lions so remarkable, and it partly accounts for their resilience under heavy persecution. They aren't as resilient as coyotes, which may increase pup production under human persecution. But heavily hunted mountain lion populations can rebound in three to seven years after heavy hunting ceases, as long as there is habitat to connect the area to other populations that can provide immigrants looking for new homes.

The "overwhelming success" Justin Clapp referred to in his presentation can be traced to decisions made by the commission seven years earlier. At that time, the commission authorized increased bag limits for mountain lions across the state in an effort to reduce the number of predators. The limits were set even higher in regions where local communities complained about livestock conflict, deer hunters complained about reduced hunting opportunities, and houndsmen sought increased hunting opportunities for themselves. Since then, Game and Fish increased limits every three years to encourage hunters to kill even more mountain lions.

The difficulty was tracking the subsequent changes in the state's dispersed mountain lion populations. Mountain lions aren't like pronghorn, hanging around in the open where they are easy to count. In fact, no state game agency knows how many mountain lions live within their borders. A few states have been bold enough to produce estimates, but most numbers are speculation. For example, Clapp relied not on numbers of mountain lions, but on three hunting metrics collected by the Large Carnivore Department within the Wyoming Game and Fish Department since 2006 to justify his interpretation that the mountain lion population had been reduced.

First, Clapp presented information on the decrease in the average age of mountain lions killed by hunters over time. Hunters generally seek older, larger mountain lions, only killing youngsters when they have no choice. Ten years earlier, Wyoming hunters killed about 40 percent subadults, but by 2016, 60 percent of the lions killed each year were young animals. Second, Clapp showed that the percentage of females killed each year had also increased over the last ten years. In general, hunters prefer male mountain lions over females because they weigh 50 to 100 percent heavier—making for a more desirable trophy. Some houndsmen also refuse to hunt females to ensure local mountain lion populations maintain their breeding pool, keeping populations healthy. Third, Clapp presented an infographic showing that the percentage of hunters who successfully killed a mountain lion each season had been decreasing over the previous ten years, even while they were granted greater opportunity because harvest limits had been increased. This was especially troubling, given that the state agency valued hunters as a tool for modulating mountain lion populations. When "interest starts to wane," explained Clapp, fewer people hunt, and hunter competition among themselves for limited prey becomes a concern as well.

In *Path of the Puma*, wildlife manager Jim Williams recounted an extreme case of hunter competition in Libby, Montana. At the invitation of the regional biologist, Williams visited Libby for the opening of mountain lion season in December of 2000. There were 115 pickup trucks in the Forest Service parking lot the day before the season opened, some with local plates and some with plates from as far away as New York—these were out-of-state hunters who had hired guides to help them hunt a mountain lion. The local hunting unit had a limit of twelve mountain lions, so clearly not everyone lined up in the parking lot was going to be able to kill one. But they were all going to try.

Within an hour of legal hunting light, the local Fish and Wildlife office had received reports that twenty-four mountain lions had been killed, double that of the unit limit, and more and more reports trickled in that first day. Local wildlife managers tried to shut it down mid-morning,

A young female mountain lion stares out from her retreat after being forced to seek refuge by pursuing hounds in northwest Wyoming. Her capture was part of research efforts in the region, and was facilitated by hounds and houndsmen Boone and Sam Smith.

but hunters were loose on the surrounding mountains, oblivious that the unit had already closed. There were just too many hunters racing to kill lions. And when the competition is high, hunters become less selective—they shoot whatever age or sex cat their hounds chase up a tree, because they are lucky to get anything at all, and if they pass up on an opportunity the unit will close before they get another chance.

Hunter competition decimated the local mountain lion population around Libby, and staff of the Montana Department of Fish, Wildlife & Parks worked frantically for a solution. Part of the problem was the proportion of tags being filled by out-of-state hunters—around half at that time—which put additional pressure on local hunters who would have preferred to spend the entire winter running their hounds but not killing a lion, rather than see the unit close so quickly. Guides were working and filling tags as fast as they could, reducing hunting opportunities for local people. In response, Montana's state agency limited out-of-state hunters to just 10 percent of the mountain lions that could be hunted each season, and followed up with a complete change in how hunting permits were authorized and managed. They dropped the old quota system and adopted a lottery system in which not all hunters who applied for hunting licenses would receive one, and those who filled a tag one year would have to wait a few years before receiving a new hunting license to kill a mountain lion again. Not everyone was supportive of the changes by any means, but enough were to initiate change. "Someone has to speak for the cats," Wade Nixon, local houndsman and proponent of change, reminds us.[3]

Wyoming was forced to follow suit in the Black Hills when similar dynamics demanded they limit out of state hunting in order to aid mountain lions and local hunters. Now southeast Idaho is grappling with the same issue, as outfitters and local hunters vie for limited hunting opportunities in which only mountain lions lose.

Wyoming is one of thirteen US states that manage mountain lion hunting. Texas is a black hole, as the state never switched mountain lions to game status, and thus they never gather any data on their abundance or trends. In Texas, people can trap or shoot mountain lions throughout the year, and they never have to report how many they kill. It's a throwback to the bounty era in terms of what people can do, only without the state providing any financial incentive for killing lions. Ignoring Texas, mountain lion hunting is permitted in every state west of Minnesota except Oklahoma, Kansas, and California. Thus, a study of harvest trends across the West, as Clapp did in Wyoming, is informative, especially given that states have been steadily increasing mountain lion hunting limits and quotas over the last ten years.

There are caveats. We don't know how many mountain lions live across the West, so we can't apply rigorous scientific techniques to determine the impact of hunting on the country's mountain lions as a whole. The best we can do is look at trends in the harvest data, which is imperfect but is arguably the best strategy in the absence of any other way forward.

It's not a clean story, either, because some harvest trends provide conflicting patterns. But overall, it appears that we are right now at a tipping point. From 2005 to 2016, there was a decline in the percentage of adult males that composed the annual harvest, and a steep increase in the percentage of subadult cats killed by hunters, both of which suggest that the US mountain lion population is declining. At the same time, there was a small decrease in the percentage of females killed each year. This generally suggests a moderately growing or stable mountain lion population in the West, although some researchers have shown that female harvest can decrease in declining populations as well.[4]

The change in the percentage of hunters who successfully killed mountain lions each season over this time period might be the most telling of all. It increased for six years, stabilized for a year, and then began to drop. This could indicate a stabilization in the population and then a slight decline. The average age of mountain lions killed across the West remained steady at 3.2 years, which is only slightly below the average age of "lightly hunted" populations, and well above that of heavily hunted populations, in which the average age tends to be about 2.5 years.[5] So overall, the outlook for mountain lions is not dire. We may not know how many of them inhabit the United States and Canada, but most experts would agree that there are probably at least 30,000.

Wildlife managers are traditionally trained in numbers meant to be plugged into what might be called economic models. They view animals

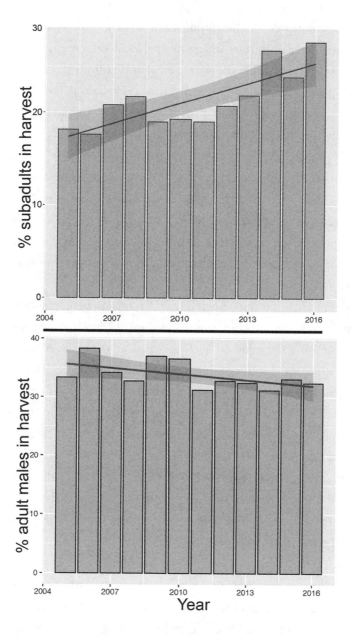

Changing harvest trends reflecting changes in the percentage of subadult and adult male mountain lions composing annual harvests from 2005 to 2016, generally indicative of a declining population. (Figure compliments of Lisanne Petracca / Panthera.)

as essentially equivalent, and females as the basic unit of breeding and production. Their objectives for game species, which are those we hunt and fish, are to maximize the number of individuals that can be harvested by people while ensuring that these populations remain sustainable for the public trust. Managers have proven time and again to be effective and proficient at maintaining and growing wildlife populations for sustainable use. When viewed from the "sustainable use" perspective, US mountain lions are doing fine. Managers watch harvest trends, and if there are indicators that the population is declining—as there are at this moment—they need only reduce hunting pressure to stabilize or grow the population again. This is called "adaptive management." Managers adapt to changes on the ground as they interpret them, and they modulate hunting accordingly.

Based on the above-mentioned harvest trends, managers should begin reducing hunting quotas for mountain lions. Otherwise, we will begin to see more obvious signs of decline—and at the scale of the mountain lion superorganism, not just a single hunting unit. The problem with focusing just on mountain lion numbers, however, is that it is not enough to decipher the health of mountain lion populations. In terms of the integrity of mountain lion populations, there is already cause for concern. Hunting pressure appears to be decreasing the number of mature male mountain lions in the West, and increasing the number of subadult mountain lions wandering around. As we discussed in earlier chapters, these changes have real consequences for us and our interactions with mountain lions.

The question remains, however, as to what removing adult male mountain lions and increasing the number of subadult animals running around the landscape does to the health and stability of mountain lion populations. Right now, we don't know, although we are beginning to gain some insights into what might be happening—it's cutting edge research today. We know well the direct effect of hunting animals, but we are just now exploring the indirect effects of hunting, which might be described as those unexpected and unintentional cascading effects of killing specific mountain lions or killing a threshold number of mountain lions on the larger mountain lion superorganism. This is the new frontier, and one that can help us move from strictly managing wildlife populations for sustainable harvest, in terms of numbers, to a management paradigm in which we ensure *healthy* mountain lion populations, in terms of maintaining their social behaviors and ecological contributions to the natural world—the world upon which we also depend.

F61 was a female mountain lion born north of Jackson, Wyoming, in July 2007, and monitored by Panthera's Teton Cougar Project. She gave birth to her first litter of kittens in 2011 when she was four years old. Using genetic techniques, Panthera scientists determined that her mate had been M21, which made sense given that he was the resident male overlapping F61's territory at the time. M21 was five years old and truly coming into his prime when he mated with F61 and her neighbor to the north, F51, at nearly the same time.

M21 was killed in April 2012 by what veterinarians were 90–95 percent certain was anticoagulant poisons—the type typically used to kill rats. He likely ingested them when he ate a smaller predator or rodent that had in turn eaten the poison, or perhaps even a carcass laced with the stuff that had been placed out as bait by anti-predator ranchers. The research team found M21 laid out on one side, legs stretched out as if he were still walking amidst the tiny green leaves of grouse whortleberry bordering a rushing river.

Several months after M21's untimely death, a young male of approximately three and a half years of age appeared to replace him. M29, as he was later named, cavorted with both F51 and F61, even though both were more than a year his senior, and both had young kittens at the time. According to most scientists, M29 should have killed M21's progeny in order to create opportunities to breed with their mothers and create his own genetic lineage. But he didn't. Instead, he fed alongside both families on various elk and deer killed by the female mountain lions.

Perhaps because his regular presence induced ovulation, or perhaps for some reason we cannot comprehend, M29 mated with both F61 and F51 while their respective kittens waited nearby. In the fall of 2012, both F51 and F61 sent their fourteen-month-old kittens packing, four months earlier than typical for females in the area. One month later, they gave birth to M29's kittens in fortified dens just six miles apart, F51's in tangled woods, and F61's in an enclave amidst moss-covered boulder scree.

M29 was killed by a hunter in October 2013, seventeen months after he established his territory; he was nearly five years old when he died, and he had likely just achieved most of his adult proportions. M29's death resulted in several consequences. For one, it was a full year before the local research team documented the presence of another male in the former territories of M29 and M21 before him. Second, it was a year and a half after his death before F61 and other females in the area were given the chance to breed again—this time with a male half their age. The absence of a resident male delayed breeding opportunities and hindered the local mountain lion population's ability to rebuild and recover.

Female mate selection is critically important to maintaining genetic health and vigor in many species, but no one has studied these obscure, indirect effects of hunting on mountain lions. Only once in her life did F61 have the opportunity to mate with a male mountain lion older than herself. In every instance thereafter she was forced to mate with three- to four-year-old males, even though she continued to age. In unhunted mountain lion populations, older experienced males hold vast territories overlapping those of four or five females. In hunted populations, mature

F61, an adult female mountain lion followed by Panthera's Teton Cougar project in north-west Wyoming. (Photograph courtesy of Neal Wight / Panthera.)

males are sometimes in short supply. What we do not know is whether F61 or the larger mountain lion population suffered for lack of oppor-tunity to choose more experienced mates.

There is one more potential consequence of M29's death. Among brown bears, also called grizzly bears, territorial males protect females with which they overlap. Their presence deters neighboring males, which would kill the cubs of resident females so they could breed with them themselves. Biologists call it infanticide when an animal kills youngsters sired by their rivals, and many mammals do it. Infanticide is a brutal form of genetic warfare, with contenders fighting for the very future of their genetic lineage in the larger population. Animals that lose are pruned from the population's genetic history—erased completely and forgotten forever. In theory, this is survival of the fittest in action—only the strongest win and thus their genes strengthen their popula-tions through the perpetuation of their genetic material. They do this by mating with as many females as possible and reducing the contributions of their rivals.

Hunting, however, throws a wrench in the entire system. Trophy hunters want the biggest and the baddest, at least when they're hunting predators like bears and mountain lions, and thus they prune strong animals from the herd.[6] When they hunt brown bears, hunters who kill large, territorial bears cause social chaos that increases infanticide for up to two years following the death of their trophy bear. Neighboring male bears shift their territories and systematically remove their competitor's progeny in their expanded territory.[7] This pattern of hunting-driven infanticide is much less clear in mountain lions as compared to brown bears, but it is suspected.[8]

Five months after M29 was killed, F51 meandered toward the eastern edge of her range, M29's two female offspring bouncing like electrons in orbit around her. They had fed off a series of elk in quick succession, and successfully dodged the local wolf pack that stole their last kill from them. F51's kittens were fat, healthy, and growing fast.

M85 was M29's neighbor to the east, and he'd begun periodically probing M29's now-vacant territory. At the time, he was on one such exploratory trip and had killed an elk in what was previously the eastern most portion of M29's range. He was lying near his kill when he heard F51 and her kittens drop through a narrow cleft in the rocks above his position. He then set forth to intercept the family. This much was clear, written in snow. What happened next involves speculation.

Perhaps M85 approached aggressively, or perhaps F51's kittens were exposed in front of her, but whatever the scenario, she engaged him. The pair met in a storm of claws and fury, flattening the snow as they rolled and wrestled. They slid down the hill about fifteen feet, packing another circle of snow and leaving behind great tufts of fur. Down they slid even further, to tumble and roll yet again, the first sprinkles of blood shining ruby-red on the untouched snow.

Then there was a last great tumble, the pair locked tooth and claw as they slid fast and slammed into a young fir tree; the lowermost branches were snapped off in the violence. Blood and fur soaked and covered the area at the base of the trunk, where they lay long enough that their entwined bodies melted into the snow. F51 died in M85's maw, orphaning two seven-month-old kittens sired by M29. M85 dragged her body twenty feet from that place before he left her. F51's kittens fled at the first signs of trouble, and M85 made no effort to pursue them. Instead, he returned to his own territory for the remainder of the winter.

The kittens survived the next few months by scavenging old carcasses they discovered buried in snow. Harsh winters kill old and infirmed elk, and this was their primary source of sustenance. But it wasn't enough. The first kitten died of starvation in June, and her sister immediately ate what remained of her. Over the next month, the skeletal remaining kitten began to be seen by tourists traveling Forest Service roads. For this reason, the state agency granted permission to local researchers to intervene, and in fact provided a road-killed animal to feed the last kitten. They fed her four times, but the damage was irreversible. Her growth was stunted by the extended starvation. She died several months later after she attacked and killed a porcupine, and its quills pierced her heart and lungs. It is certainly possible that killing M29 led to the death of F51 and her offspring as well—one bullet, four mountain lions. This is the sort of indirect effect of hunting very difficult to capture and describe, and in fact, we lack for research that investigates these connections.

The direct effect of hunting mountain lions with hounds is the removal of older mountain lions, often males, and a proportion of other sex and age classes. The indirect effects of hunting mountain lions are the resulting skewing of both male–female ratios and also the proportional representation of age classes. In hunted populations, for example, male mountain lions' territories overlap more in space, increasing the potential for physical fighting and conflicts that hurt mountain lions.[9] The consequences of these changes for us have been discussed in the preceding chapters; this is social-chaos theory or juvenile-delinquent theory at work. The consequences for mountain lions, however, are little understood.

Mountain lions, it turns out, are not all the same. This is not surprising, given their curiosity and intelligence, but it still warrants emphasizing. The traditional approach to wildlife management assumes that animals are equivalent, and it's just not true. Individual mountain lions eat slightly different diets. Some are social, others less so. Some females are

A kitten lounges under the care of her attentive mother. Not all mothers are equally proficient at defending their young and teaching them to survive.

excellent mothers who successfully raise numerous kittens, while others are unable to do so. How is a mountain lion population affected when a productive female is killed versus one that is a poor mother incapable of raising young? These are questions that need answering.

Mountain lions feel pain. They exhibit fear and affection, and their family units form close bonds for the duration they remain together. These facts challenge the current approach to wildlife management, which is essentially regulated harvest that prioritizes production over the welfare of individual animals. Mountain lions are complex animals that live in interconnected communities. They interact with other adult mountain lions with regularity and perhaps predictability.[10] Food appears to be a social currency, bonding mountain lions together through reciprocal food-sharing. Some biologists dislike the word *sharing* because it implies motivations we cannot know. They prefer to say that mountain lions exhibit tolerance of other mountain lions at their kills. The reciprocal patterns of this tolerance, however, implies some cognitive understanding and investment in relationships; it also implies that mountain lions rely upon their communities to survive.

Mountain lion reciprocity is not random, and it occurs within clusters that might be described as "neighborhoods." Mountain lions in the same neighborhood hunt the same herds, wander the same paths, and allow their kittens to play on the same log fortresses; they socialize and share food with other members of their own neighborhood far more than they do with mountain lions in adjacent neighborhoods. But mountain lion neighborhoods, unlike our own, are delineated by the physical boundaries of male territories. Males function like mayors or governors of fiefdoms that influence which mountain lions interact most frequently with each other. Thus the removal of a territorial male may be more disruptive to social dynamics than removing a female. But, at this time, we simply don't know.

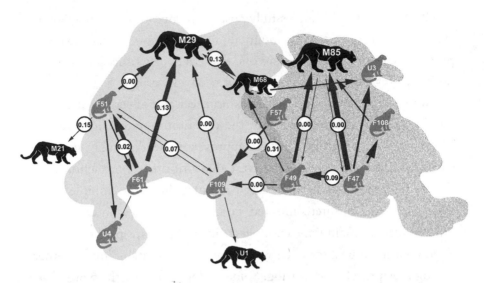

A conceptual representation of two adult male home ranges (M29 and M85, represented by black icons) in northwest Wyoming, with their respective overlapping female mountain lions (gray icons). The arrows represent actual food-sharing that occurred between individuals, and the direction indicates which animal shared with which. The width of the arrow reflects the frequency of food-sharing—the thicker the arrow, the higher number of carcasses shared. The numbers in circles reflect genetic relatedness between the two individuals on a scale of 0–1, highlighting that food-sharing occurs predominantly among unrelated animals.

As managed in most of the western United States and Canada, mountain lion hunting is not a threat to the long-term sustainability of mountain lion populations. This speaks specifically to the direct effects of hunting on mountain lion numbers, and it can't be said any more plainly than that. Mountain lion hunting, however, is almost certainly the greatest threat to the *integrity* of mountain lion populations. This speaks to the differences between the direct and indirect effects of hunting.

Ultimately, we need to shift the discussion of mountain lion hunting from one of numbers and "sustainable harvest" to one in which we strive to maintain healthy, ecologically functioning mountain lion populations on the landscape. Step one, then, is to determine what a healthy mountain lion population looks like. For twenty years, Ken Logan, one of the people who has studied mountain lions longest and most intensively, has been suggesting that each state set aside large areas where mountain lions are protected in order to allow mountain lions to begin to illustrate natural biology.[11] Logan continues to stress how important these areas are for science because of our need for "reference" populations with which to compare predation rates, density, and so many other things exhibited by lightly and heavily hunted populations.[12] At this time, zero states that allow mountain lion hunting have created such sanctuaries.

Even after sixty years of intensive research in North America, there is still so much we don't know about mountain lions. How many mountain lions were here, for example, roaming North America before the Europeans arrived? No one knows. We need a time machine to send biologists back 500 years to see how mountain lions lived in primordial North America, alongside wolves, jaguars, grizzly bears, and Native people. Such knowledge would be invaluable toward understanding what a healthy mountain lion population, or at least a natural one, might look like.

We do not know the carrying capacity for mountain lion populations in any given area—if we knew how many mountain lions an area could support, we could compare current abundances with this benchmark as one metric for assessing the health of a given mountain lion population. Recent mountain lion studies have provided widely varying estimates of their abundance, and in essentially all of them, the mountain lion population suffered some sort of human influence. The reality is that there are few places left in all of North America where we can study a natural population unimpacted by humans. But even if a protected place existed where humans were allowed little sway, it would still unlikely be reflective of historic mountain lion populations that had to compete with a suite of other large carnivores.

Natural systems are not egalitarian. Dominant species win in contests for resources and physical fights. Occasionally winners eat losers. In North America, mountain lions are subordinate to black bears, brown bears, jaguars, and gray wolves.[13] The presence of wolves, in particular, profoundly impacts mountain lion abundance and behavior. In northwest Wyoming, for example, the effects of just twenty wolves is equivalent to local human hunting pressure on the same mountain lion population.[14] Wolves kill mountain lion kittens, and chase and harass mountain lions of every age. They steal their food, and displace mountain lions from the best habitats with the most prey. It may be that mountain lions were never abundant across North America before

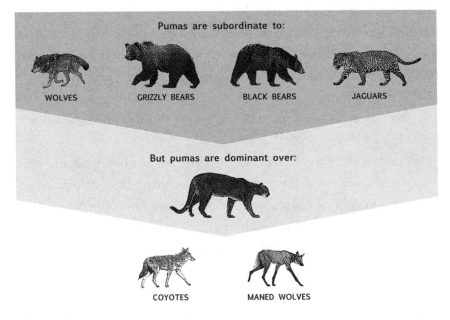

Mountain lions are apex predators that are subordinate to other more dominant predators, and this fact influences diverse aspects of their ecology. (Figure created in collaboration with Danielle Garbouchian / Panthera.)

European settlement, because everywhere they lived, they lived with wolves. We just do not know.

What we have accomplished in the management of mountain lions, given the tremendous uncertainty in our knowledge about the effects of hunting on mountain lions and what a natural mountain lion population might look like, is miraculous. Part of our success, if we are bold enough to call it that, is due, no doubt, to the cat's own resilience. Part of it may also be the absence of wolves over most of the US. Mountain lion management is a bit like groping in the dark in an unfamiliar room. That said, state managers have done well at maintaining mountain lion numbers, even under heavy pressure to hunt them hard in order to protect ungulates and livestock. A common argument for their success is that the US mountain lion population is still growing. In contrast to the harvest trends we discussed above, many point to increasing reports of mountain lions in the Midwest as evidence that the population is growing. The assumption of state managers is that eastern migrants are extra mountain lions that couldn't find suitable range in an overcrowded West, so they dispersed eastward into lands without mountain lions instead.

The counterargument is that these rogue mountain lions in the East are descendants of mountain lions that survived the onslaught brought on by European settlement. State agencies across Midwestern and Eastern states record thousands of mountain lion sightings every year, and some of them are verified with field evidence. So, let's briefly cast our eyes east of states that manage mountain lions into lands where mountain lions once roamed. Can we determine whether mountain lions are expanding eastward to reclaim former range, or have they been lurking in the shadows all along, awaiting people to make their presence known?

CHAPTER 6
Lions on the Eastern Seaboard

On January 23, 2018, the US Fish and Wildlife Service (USFWS) released a new rule to remove the "eastern cougar" from the federal list of endangered and threatened wildlife. The report began, "We, the U.S. Fish and Wildlife Service, determine the eastern puma (=cougar) (*Puma (=Felis) concolor couguar*) to be extinct, based on the best available scientific and commercial information." What followed was sensationalist media coverage about the supposed extinction of a cryptic species and angry citizens certain there were still mountain lions prowling Eastern states. Conspiracy theories circulated claiming federal agents denied big cat sightings east of the Rocky Mountains in an attempt to stymy storytelling in rural watering holes across the Eastern Seaboard.

As it happened, in 2011 a young male mountain lion was killed on the Wilbur Cross Parkway in Milford, Connecticut. The animal's carcass provided irrefutable evidence that mountain lions inhabited New England. How then could the USFWS claim that eastern mountain lions were extinct so quickly thereafter? It was the sort of logical inconsistency that fueled the concerns of many who live in the eastern United States and Canada, and accuse Eastern wildlife agencies of an "institutional resistance to the existence of pumas."[1] We will never know why the USFWS emphasized the word *extinction* at the start, given that it had to know few would read the full fourteen-page report. In a world of tweets and social media posts, people look for sound bites, not detailed accounting. In this case, however, the devil was in the details.

We know that mountain lions existed in the East when European settlers landed their boats. We are 100 percent certain that early settlers waged comprehensive war on predators, and we've been told that they successfully wiped out mountain lions along the Eastern Seaboard. The last legitimate sighting of a mountain lion in New Hampshire was in 1885, for example. So, the question remains: Is it possible that a few mountain lions—animals expert in stealth and camouflage—survived

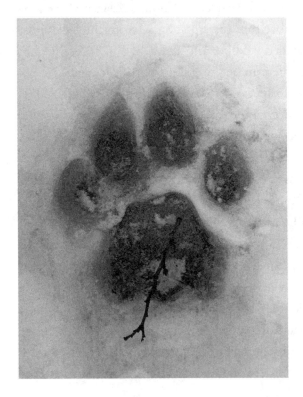

The slightly melted hind footprint of a female mountain lion in snow. Mountain lions, when present, leave ample evidence of their existence: footprints, carcasses of prey, latrine sites, and various signs of scent marking.

among the darkest, densest woods of the Eastern forests, and that people occasionally glimpse these animals in backwoods haunts? It's a loaded question, as in fact, mountain lions did survive completely undetected for many decades along the Eastern Seaboard in southernmost Florida swamplands. The last of them may have disappeared completely unbeknownst to the bustling USA, if not for private investments that paid for concerted efforts to look for them. Florida aside, however, there is no evidence that mountain lions survived the onslaught east of the Mississippi River. There were few mountain lions that survived anywhere east of the Rocky Mountain range following the bounty years. So why then, do people keep claiming to see them?

John Lutz is among the most vociferous of those who believe that mountain lions were never eradicated in the East. He believes that native big cats still live in most, if not all, Eastern states. His LinkedIn profile describes his current work as "professional cougar-puma-mountain lion researcher" for the Eastern Puma Research Network, which he founded himself and has worked with since 1963. He's a passionate man who believes that the "Deep State" is covering up eastern mountain lions for the dubious reasons of hoodwinking their constituents. He is also a man who wants to see mountain lions proliferate and be protected across the East.

State and federal agencies "keep denying, denying, denying the presence of wild, free-roaming cougars" in the East, Lutz explains.[2] Since 1965, he has gathered more than 12,000 records of mountain lions in Eastern states, which he carefully organizes in three-ring binders that he maintains at his residence. He estimates that ten to twenty big cats live in the Tennessee portion of the Smoky Mountains, and fifty to sixty cats roam upstate New York.[3] When officials declared that the mountain lion killed in Connecticut in 2011 originated in South Dakota, Lutz countered that it was another government cover-up and that the cat had actually originated in eastern Canada.[4]

Mark McCollough is an endangered-species biologist with the US Fish and Wildlife Service, and among his responsibilities over the years was leading a five-year investigation into the status of the eastern cougar, including a thorough review of existing evidence of recent mountain lions in the East. Like Lutz's Eastern Puma Research Network, the USFWS receives thousands of mountain lion reports from residents in Eastern states, and they continue to accumulate to this day. In some areas of the East, up to 30 percent of sightings received by USFWS are of black mountain lions, the existence of which we debunked in chapter 1.[5]

McCollough's 110-page review was thorough and succinct. He found zero evidence of native mountain lions living in the East.[6] Most reports, he concluded, were cases of misidentification. Reports of mountain lion footprints were mostly dog tracks, and most photographs of supposed mountain lions were Labrador retrievers, bobcats, coyotes, or house cats that looked larger than they were until one deciphered the scale of objects adjacent to the animal. That is not to say that McCollough didn't discover evidence of mountain lions, because he did. The report included 110 verified mountain lion records in the East. The difference between McCollough's and Lutz's conclusions, however, was that every one of these records could be explained by released pets and trade animals from distant origins. As of 2005, there were an estimated 110–135 legal, licensed mountain lions held by exotic pet owners and zoos in the Northeast.

Neither did McCollough find evidence of native breeding mountain lions. Someone shot a mountain lion kitten in New York in 1993, and another kitten was struck by a vehicle in Kentucky in 1997, but genetic tests revealed that both animals originated in South America. Genetic tools, in fact, proved to be a powerful means of distinguishing fact from fiction in McCollough's review. They indicated that about a third of eastern mountain lions for which they had scats, hair, or tissue were actually from Central or South America. In conclusion, the USFWS review provided clear evidence that mountain lions, at least on occasion, do wander through eastern states. Therefore, McCollough's report vindicated those who claimed they'd seen mountain lions. Some people really see big tawny cats in the East, or encounter their footprints in the snow. Nevertheless, the report also stood counter to the notion that rare mountain lions found in the east were native to the area, or that some remnant population had miraculously survived, invisibly, among millions of people for 100 years. Every mountain lion recently found in the East was, in fact, not from the East at all.

Utilizing genetic tools, scientists were also able to determine that the 2011 Connecticut mountain lion originated in the Black Hills of South Dakota. Thus, the debate became not whether it was an eastern cougar, but whether some overzealous, but misguided individual had driven him east and released him in New England in an attempt to initiate a new breeding population. The wonders of genetic tools provided remarkable illumination into this question as well. Scientists were able to compare the Connecticut lion's genetic finger print with other individual profiles collected from fur and scats across the Midwest over the previous several years. Scientists confirmed that the Connecticut mountain lion was detected at an additional five locations in Wisconsin, Minnesota and New York during its eastward migration and before its demise.[7] Investigative journalist Will Stolzenburg went further in his book *Heart of a Lion: A Lone Cat's Walk across America*, in which he pieced together scats, footprints, and photographs captured by motion-triggered trail cameras to mark thirteen distinct places where the Connecticut lion trod between the Black Hills and its final resting place on the Wilbur Cross Parkway. These revelations astounded even the most stoic scientists. This young male mountain lion walked something like 1,700 miles over as much as two years before he was killed by a speeding SUV. It is the longest dispersal ever recorded for mountain lions anywhere in their range.

Right now, mountain lions are moving east out of western strongholds more often than ever before. From 1990 to 2004, twenty-five mountain lions were killed on roads or by other means across the Midwest.

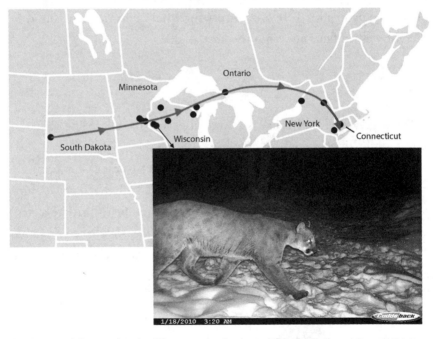

The potential dispersal path of the young male mountain lion killed on a Connecticut highway in 2011, based on evidence collected by Stolzenburg, *Heart of a Lion*, and Hawley et al., "Long-Distance Dispersal of a Subadult Male Cougar from South Dakota to Connecticut Documented with DNA Evidence." The inset is a photograph of the actual lion photographed during his dispersal across Wisconsin in 2010. (Photograph courtesy of Lue and Krystal Vang.)

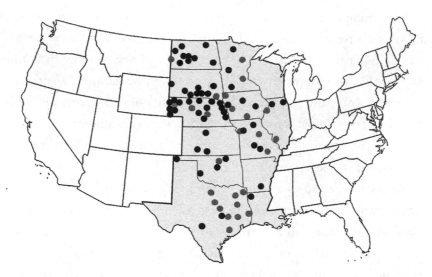

A map of counties in which mountain lion carcasses were collected across the Midwest from 1990 to 2015. Carcasses are irrefutable evidence of mountain lion presence. The gray circles represent counties in which carcasses were collected from 1990 to 2004, and the black circles represent counties in which carcasses were collected from 2005 to 2015. Note that circles represent one or more carcasses found in the same county. Figure modified and re-created from one originally published in Larue, Nielsen, and Pease, "Increases in Midwestern Cougars Despite Harvest in a Source Population," Fig. 1.

From 2005 to 2015, that number soared to 86.[8] As we discussed at the close of the last chapter, many argue that this is evidence of a growing mountain lion population in the West, and that the "extras" unable to find space amidst existing mountain lion communities are seeking territories of their own.[9] Alternatively, some animals exhibit higher dispersal rates under persecution, and so the greater number of mountain lions in the East may instead be evidence of overhunting mountain lions in the West. Or it may be that eastern mountain lion dispersal patterns are unrelated to western mountain lion densities or hunting pressure altogether. Regardless of the reason, mountain lions are definitely moving east, albeit at a sluggish pace. We shouldn't expect to see kittens scampering along the Eastern Seaboard anytime soon.

Male mountain lions, as discussed in previous chapters, almost always disperse from their natal range. Males also tend to disperse farther than females, and thus males will move east first, forging the paths that females will follow to establish breeding populations. Wandering males seek company, not prey. There are plenty of deer across the Midwest, but few companions. So, their dispersals can be prolonged. The Connecticut male is a perfect example of a lone male wandering an incredible distance in search of community.

Unless a male moves east in the company of a female, or he finds a female awaiting his arrival in the East, there isn't an opportunity for a breeding population to become established. For this reason, breeding mountain lion populations are expected to expand eastward at the rate of female expansion, not male, and to occur in fits and starts based upon the availability of suitable islands of safe habitat and human tolerance for predators that have been absent in the Midwest for a century or more. It took roughly twenty years for dispersing mountain lions to successfully leave the Black Hills of South Dakota and cross 100 miles of tamed habitat to establish a breeding population in the wilder Pine Ridge area of Nebraska.[10] From there it's about 1,600 miles to the Atlantic Ocean, or more than 300 years of expansion at the same pace.

Eastern migrants, however, complicate the story of eastern mountain lions. Wild mountain lions that reach New England on their own are still not eastern cougars. In parallel to McCollough's report emphasizing the fact that recent mountain lions in the Northeast were not native to the Northeast, the same is true for eastern migrants. The protections granted the eastern cougar under the Endangered Species Act were only for those mountain lions native to the northeastern United States and Canada. Western mountain lions that travel east have been and continue to be protected, following the laws of the various states through which they move. In general, mountain lions are already protected across the East by state legislation.

What is perhaps most confusing about the opening to the federal ruling declaring the extinction of the eastern cougar is that there never was an eastern cougar. The entire listing of the eastern cougar in 1973 was a case of mistaken identity, based on the false assumption that mountain lions in the northeastern United States were a different subspecies than those in other parts of the country. The newest science, however, has revealed that the classification of an "eastern cougar" subspecies was never warranted in the first place. The USFWS ruling does include this discussion, if you dig deep.

Phylogeny is the science that proposes how animals are related to one another, based upon their shared evolutionary history. In centuries past, phylogeny was based upon the shape, color, and external characteristics of animals and their skeletons. Early scientists described thirty-two distinct mountain lion subspecies across North and South America, the eastern cougar among them, corroborated by differences in the length and width of their skulls and the subtle color variations of their coats. In contrast, when we apply modern techniques and look at the DNA of all these cats, we see a very different picture.

The International Union for Conservation of Nature (IUCN) currently recognizes six subspecies of mountain lions, of which only one— *Puma concolor cougar*—inhabits all of North America, excepting the southernmost portion of Central America north of the Panama isthmus. In other words, the eastern cougar was never a separate subspecies, and mountain lions that historically inhabited the northeast of North America were the same subspecies as the mountain lions in Idaho and Montana, the descendants of which are still very much alive today. Thus, the eastern cougar is not extinct—it never existed. It is more correct to say that the North American subspecies of mountain lions is locally extinct in the Northeast, because there is no breeding population in this region.

A potentially interesting consequence of the USFWS ruling is that it may have paved the way for more easily reintroducing mountain lions in the East. Reintroducing animals into their historic range, as wolves were in Yellowstone National Park and central Idaho in 1995 and 1996, is an arduous process. Reintroducing federally protected animals is even more difficult, because they are so closely regulated. In contrast, it's much easier to undertake the reintroduction of an unprotected species. Even better, people can now propose to reintroduce the *native* mountain lion subspecies to New England rather than replace it with a different one from the West. This is a much stronger ecological argument, and it undermines any opposition against introducing nonnative animals that do not belong. Lutz and the Eastern Puma Research Network may continue to claim that western mountain lions are larger, more aggressive and different than eastern mountain lions, but there is absolutely zero evidence to support them.

Remarkably, there is a small breeding population of mountain lions that resides east of the Mississippi River. Often overlooked because they aren't called cougars or mountain lions, Florida panthers miraculously survived European settlement amidst the protection provided by clouds of mosquitos, sharp-toothed alligators, and sucking, stinky swamplands difficult to develop into shopping malls and condominiums for retirees. The USFWS listed the Florida panther as endangered in 1967, but Florida officials believed that the cats were already extinct and pushed back against the potential impacts of the Endangered Species Act (ESA) on new development and wildlife management. Then the World Wildlife Fund (WWF) stepped in to stop any drastic declarations that panthers were extinct.

In 1972, the WWF hired a Texas lion hunter named Roy McBride to search for elusive panthers. McBride and his band of hunting hounds traversed the southern swamps and indeed, they found panther footprints and scats, but they never laid eyes on an actual cat. He returned in 1973, and his hounds sniffed out and chased a withered old female panther up a tree near Fisheating Creek, southwest of Lake Okeechobee.[11] The cat was in rough shape, skinny and tick-ridden, but she was alive—a real mountain lion living in southern Florida. McBride found several others in quick succession, and agencies quickly worked together to protect them. The state of Florida listed the panther as endangered that very year.

Florida's early conundrum was not too many but too few mountain lions. In fact, there were so few animals back in the 1980s and 1990s that mother panthers were breeding with their sons, brothers were mating with sisters, and inbreeding was resulting in unusual angular facial features, deformed tails, and, most catastrophically, in offspring unable to propagate themselves. Following great debate, Roy McBride was dispatched back to Texas to catch eight female mountain lions. They were carted to Florida in 1995 and released in pairs in four different areas across southern Florida in an attempt to spread their genes. It was a genetic "rescue" mission, one that many argued was a waste of money and effort. But twenty years later we know it was a brilliant gambit and an astounding conservation success story.

Nevertheless, those charged with protecting panthers also failed them. The ESA should have safeguarded the landscapes needed to sustain panthers, but questionable science and political corruption allowed for the approval of thirty-five development projects that permanently destroyed 38,484 acres of actual and potential panther habitat.[12] If there's nowhere for them to live, it doesn't matter how many panthers we save. Between 2000 and 2003, a whistleblower revealed the politics

The release of a Florida panther back into the wild. Panthers that survive being struck by vehicles are rehabilitated and released to ensure that as many individuals as possible can contribute to the population's persistence. (Photograph by Tim Donovan / Florida Fish and Wildlife. Used under Creative Commons license.)

guiding the approval of new development in panther habitat, and a new panther working group composed of outside scientists identified flaws in previous research that had supported the political maneuvering. Since that time, proposed development near panthers has been scrutinized. Unfortunately, significant damage had already been done. Today as many as 200 panthers wander southern Florida, but every year, approximately twenty-five panthers are killed on roads, and more die from other causes. Further genetic rescue missions are needed to maintain healthy panther populations over time.[13]

The 2018 eastern cougar ruling did not impact Florida panther protections, at least not directly. Panthers remain listed as endangered by both the US Fish and Wildlife Service and the State of Florida. As mandated by the ESA, federal biologists created a recovery plan for the species, dictating the circumstances under which panthers could be removed from the endangered species list. In Florida, recovery objectives require two populations of panthers in historic range, each composed of 240 independent animals and each sustained over a minimum of twelve years. At this point there is only a single population, and it is below target numbers. Panthers are unlikely to be removed from the endangered species list anytime soon.

The caveat, however, is that our understanding of current phylogenetics no longer supports the Florida panther subspecies. Just like the eastern cougar, Florida panthers are the same subspecies of mountain lions found across the West. But because it would be devastating for Florida panthers to be removed from the endangered species list, many speculate that the USFWS may switch the basis for listing the Florida panther to one of a "distinct population segment" (DPS). Such a justification is defensible because Florida panthers are separate from every other breeding population of mountain lions, and thus particularly vulnerable to extinction. They cannot be rescued by dispersers in the larger mountain lion superorganism discussed in chapter 5, for example. The second criterion for DPS listing is that the population is significant, meaning that it is important in some way. As the only breeding population of mountain lions east of the Mississippi River, panthers are invaluable to biodiversity. Only time will tell if federal officials who review the status of the Florida panther agree.

Mountain lions are undeniably moving eastward out of their current strongholds in the West. Nebraska, for example, was quick to commence legal hunting of its tiny population of mountain lions. At last estimate, Nebraska's largest of two breeding populations was sixty animals. It's fair to ask whether it was fear of large cats or a desire to provide local hunters new opportunities that drove Nebraska officials to begin hunting so quickly, as there is no biological justification for hunting a population so small and already suffering other forms of mortality like vehicle strikes and poaching. It's only a matter of time before every Midwestern state is wrestling with whether to establish a sport hunt for mountain lions, or to fully protect the species as California did. Should a state decide to initiate hunting, we can only hope that they allow mountain lion populations to grow large enough first, in order to ensure steady eastern recolonization of their historic range.

More than anything, the uptick in eastern migrants and the speed with which they are killed signals the need for sweeping education about mountain lions, or perhaps a positive public campaign rolled out like a red carpet before them. Pioneering mountain lions wander into agricultural landscapes that haven't supported large carnivores for a century or more, and not everyone is excited to see them return. Changing people's beliefs about large carnivores, as we'll discuss in the next chapter, is challenging. If mountain lions successfully navigate the soy fields, cornfields, and dairy farms of the Midwest, they'll enter the denser suburban and urban landscapes of the East, with their spiderweb of roads supporting substantially more lethal, speeding vehicles. The eastern United States supports more people, more roads, and more cars than the West, and lacks the large public lands in which it's easier for mountain lions to disappear. Without doubt, mountain lions that succeed in establishing themselves in the East will find life difficult. But given the chance, they will return. They are already on their way.

CHAPTER 7
How to Love a Keystone Predator

Josh Barry effortlessly hiked the steep slope, a sturdy young man with a physique that reflected a lot of time lifting weights. His pack bulged in odd places, and attached to its side were mysterious strips of aluminum siding that glinted in the sun, strikingly out of place in a wilderness setting in northwest Wyoming. He ascended an open hillside dotted with rocks and sparse vegetation, crested the ridge above, and dropped into a fold in the landscape that protected a lush, forested bowl at high elevation. The understory was knee-high, green grass interspersed with purple larkspurs and scarlet gilia. The overstory was quaking aspen that tinted the light a pleasant shade of green.

"We're getting close," he said so softly that he might have been speaking to himself. "I can smell it."

He continued through the aspen grove to its far side, where the slope began anew but was crisscrossed by the fallen trunks of dead fir trees. Strewn across several logs was a bloated elk carcass, which, with Barry's arrival, burst into an angry cloud of buzzing flies. He quickly assessed the area, noting the maggots already at work on the carcass, and the space limitations imposed by the lattice of logs. When he spotted beetle larvae jostling for position on the massive carcass, he broke into a smile. Josh was in fact, looking for beetles.

He set about clearing three areas around the dead elk, digging paired, shallow holes exactly three feet apart. In each, he placed plastic cups filled about a third of the way full with water and dish soap. He placed the aluminum strips in grooves that he dug between the cups, forming a barrier that would guide beetles on the ground. The idea was simple. Beetles attracted to the carcass fly in from afar. They land to walk a short distance to the carcass, and run into the barrier. They follow it one way or the other until they drop into a cup, where the dish soap destroys the viscosity of water and they drown. Barry built three such barriers around the carcass, allowing for equal space between them for some beetles to avoid the traps and reach the carcass unmolested. He jotted a few notes about the location and state of the carcass in his notebook, packed up his tools, and then retraced his footsteps down the mountain. The flies returned *en masse*.

Barry was a graduate student researcher with Pace University, where he received training in environmental science and statistics. His fieldwork was with Panthera's Teton Cougar Project in northwest Wyoming, where the team taught him about mountain lions, conservation in the real world, and how to cook over open flame. The focus of Barry's work was determining which beetle species utilized carcasses of animals killed by mountain lions. It was an obscure area of inquiry, but it made sense within the larger framework of Panthera's Teton Cougar Project. One aspect of Panthera's larger research strategy was filling holes in our understanding of the ecological roles that mountain lions play in natural systems. More to the point, Panthera operated under the assumption that sharing the positive ways in which mountain lions benefit ecosystems that we humans also depend on will increase people's appreciation for mountain lions more broadly. It's not as crazy an idea as it sounds, as recent research by sociologists have highlighted that one of the best strategies for changing people's values about large carnivores is in fact this very approach—to highlight the contributions that large carnivores make to their environments and human society.[1]

Loosely, wildlife *tolerance* is defined as the "passive acceptance" of animals.[2] Most people generally accept moose, for example—the big chocolate-colored animals that live in snowy regions—for what they are, and harbor no ill will toward them. In other words, most people *tolerate* moose. In contrast, some people do not tolerate mountain lions. They harbor negative attitudes about them, or believe they need to be controlled, excluded, or exterminated.

Today, large predators like mountain lions and brown bears live only where we permit them to do so. Humans impose a *social carrying capacity* upon mountain lions and other large carnivores, very different than their biological carrying capacity as described in chapter 4. How many mountain lions are people willing to glimpse in a year or a lifetime? How

A female mountain lion and her seventeen-month-old "teenage" male offspring, already significantly larger than herself, sharing affection to reinforce their bonds. One approach to building human social capital for mountain lions might be described as "pulling on heart strings" by revealing their affectionate family lives. (Photographs courtesy of Jeff Hogan / Hogan Films.)

many dogs and sheep are they willing to lose before it's just too much predation? These are the sorts of questions that place threshold limits on the social carrying capacity that a community or culture imposes on mountain lion abundance. Intolerance and, more importantly, illegal killing because of intolerance are among the most influential factors in determining whether a carnivore population declines, remains stable, or thrives and grows. Therefore, improving tolerance for mountain lions is a critical first step toward living with them, rather than against them. It is also an essential strategy toward seeing them recolonize eastern haunts where they have been absent for so long.

Traditionally, people invested in increasing tolerance for mountain lions tried to mitigate people's fear of them. Advocates incorrectly assumed (1) that fear and tolerance are inversely correlated—less fear means greater tolerance, and (2) fear is rational—so better information reduces fear. For example, when we tell people that they place themselves at greater risk of injury when shaking vending machines to loosen candy bars that get stuck in their descent than they do when hiking in mountain lion country, we do not necessarily increase their tolerance of mountain lions (though it may increase their fear of vending machines). This is because lambasting people with these kinds of statistics doesn't reduce their fear of large carnivores. It might even remind them that the risk is real, even if infinitesimal, and therefore *decrease* tolerance for large carnivores as people become more afraid. Every well-intentioned mountain lion book rattles out statistics about mountain lion attacks with the intention that their rarity will teach people to control their fear. Recent research on tolerance, however, suggests that this approach may in fact be part of the problem. So what should we be doing instead?

As it so happens, there's been a tremendous amount of recent research to determine what influences people's "tolerance" for wildlife, as well as what might be the best tactic for improving tolerance for large predators. Part of the solution lies in a better understanding of the animal,

as discussed in chapter 1. Another part lies in reducing negative interactions between mountain lions and people, as discussed in chapters 2
and 3. And another lies in sifting fact from fiction, which is the theme
of this book. There are, however, several additional strategies that conservation practitioners employ to try to improve tolerance for wildlife.
They might broadly be categorized as strategies that respectively speak
to peoples' wallets, hearts, and minds.

There's nothing more terrifying than flying along some motorway when,
out of nowhere, a deer leaps into the path of the vehicle. Some people
slam on the brakes, some swerve, and some collide. The options are generally split-second choices between bad and worse. In the United States
alone, there are 1.2 million collisions with deer every year, resulting in
about 200 human fatalities, 29,000 human injuries, and nearly $1.7 bil-
lion in damages to vehicles, private property, and other infrastructure.[3]
Not to mention more than a million dead and injured deer.

People are much more likely to be killed by a deer or because of a
deer than by any other wild animal, especially in the East, where deer
numbers have exploded due to the abundance of good food in suburban yards and the absence of wolves and mountain lions that would
have at least diminished their numbers. Deer cost people in other ways
as well. They damage gardens, food crops, and forestry industries, and
they spread diseases to other species we value, like elk and moose. They
hoover up acorns and other mast, impacting tree recruitment that builds
forests as well as the lives of other animals that rely upon mast resources.
Mountain lions, argue Dr. Sophie Gilbert and colleagues, are part of
the solution already in place in the West, and they could be in the East
as well, should they colonize on their own or be introduced by state
managers.

One popular approach to increasing tolerance for large predators is to speak to people's wallet–mind connection rather than their hearts. *Ecosystem services* are benefits provided by wildlife and other nature that specifically benefit people. Ideally, these services can be quantified to emphasize their importance. Brazilian free-tailed bats, for example, provide millions of dollars in pest control to the agricultural sector each year, both by reducing the need for more pesticides and by reducing pest damage across the United States and Mexico.[4] Globally, animals provide more than $200 billion in pollination services to maintain and grow the most important food crops that feed the human population.[5] The general idea behind reporting ecosystem services is that everyday people unconcerned about wildlife perk up when they hear about their economic value; subsequently, they increase their tolerance for wildlife and begin to appreciate the animals around them.

The ecosystem services strategy has only rarely been used for mountain lions. The best example is that presented by Dr. Sophie Gilbert and her team. They not only estimated the economic value of services provided by mountain lions by reducing deer–vehicle collisions in dollars, they also measured it in human lives. To accomplish this, they completed two different analyses. First, they calculated the economic contributions of the nascent population of mountain lions in the Black Hills of South Dakota, by comparing the number of deer–vehicle strikes before and after mountain lions re-established a breeding population. Once in place, resident mountain lions reduced deer collisions by 9 percent, which translates to an annual savings of approximately $1.1 million for the people of South Dakota. That's an impressive savings.

Second, Gilbert's team conducted a predictive calculation of what mountain lions could save people should they re-establish in the East. They predicted that mountain lions would reduce deer numbers by 22 percent, after which the deer density would stabilize again at a reduced (and healthier) number. Over thirty years, the effects of re-established mountain lions on deer populations would result in 21,400 fewer injuries to people and 155 fewer human fatalities, and it would avert $2.13 billion spent in damages and reparation. Contemplating these figures makes one's head spin.

In another ecosystem services study, researchers followed nine mountain lions in Colorado and discovered that they were disproportionately killing deer infected with chronic wasting disease (CWD).[6] This is a debilitating and contagious neurological disease in which animals suffer slow deterioration of the brain, resulting in abnormal behavior, lethargy, reduced awareness of one's environment, starvation, and death. Game managers are particularly worried about CWD as it spreads from deer to elk and moose. The only way to stop, or at least slow, the spread of CWD is to hire sharpshooters to prune sick deer from the herd. If it's true that mountain lions target CWD-infected deer (and we'd need to replicate the study with larger sample sizes to be sure), mountain lions could save state agencies the costs of hiring sharpshooters, and, more importantly, save untold thousands of deer, elk, and moose from terrible deaths. This is something even deer and elk hunters can appreciate.

Others have speculated that reintroducing mountain lions in the East could reduce Lyme disease, a debilitating human disease sweeping the Eastern Seaboard and moving west, but this is pure speculation at this point. In fact, researchers have proposed that coyotes, red foxes, and other predators that specialize on small mammals may be more important than mountain lions in controlling Lyme prevalence and spread.[7]

The counterargument to citing ecosystem services as a conservation strategy to convince people that large predators are good is that it may backfire for species for which we lack information. The ecosystem services argument lays the burden upon biologists to get their act together and document all the positive roles animals play in natural systems so that they can then collaborate with economists to estimate each species' net value. For mountain lions, most research is dictated by state agencies with no interest in permitting or funding this sort of study, and so it's unlikely we'll have a repertoire of ecosystem services research upon which to draw.

Keeping with the theme of speaking to people's wallets, some researchers and conservation organizations bypass the ecological research and go straight to providing a financial incentive to people who conserve mountain lions and other predators. For example, the Northern Jaguar Project in Sonora, Mexico, with support from the American NGO Defenders of Wildlife, pays ranchers for photographs of carnivores captured on their properties: $300 for a jaguar, $150 for an ocelot, and $100 for a mountain lion.[8] This is real money in that part of Mexico, and it provides incentive to ranchers to keep animals alive. Carnivores can also continue to pay out over time as they are photographed again and again. In addition, such a system provides an incentive for ranchers to see wild game flourish on their land, which might in turn attract more wild cats, generating additional income. The problem becomes in ensuring that the pot of money never runs dry—what happens when the ranchers aren't paid to maintain carnivores on their lands? It's referred to as the "white elephant" phenomenon in some circles when an NGO with the best of intentions invests in a community for a few years. Then the money dries up, after which people return to the way they were living prior to the "intervention." In contrast, though, some ranchers come to value the photos of rare species more than the cash incentive—they are trophies of a sort, and many ranchers feel pride in knowing they support healthy ecosystems.

Nevertheless, the heart of the counterargument to ecosystem services and other wallet-based approaches is that they may discourage people from recognizing the intrinsic value of wildlife above and beyond what they can do for us. Isn't it a better approach to improving tolerance just to get people to fall in love with mountain lions, rather than to invest the money and time needed to estimate their economic value to human communities and healthy ecosystems?

Wolves made their comeback debut on the open expanses of the National Elk Refuge north of Jackson, Wyoming, at the start of February 1999. Absent since 1926, wolves were reintroduced into Yellowstone National Park in 1995 and 1996. Several years later, the first wolves trickled south to Grand Teton National Park, about 120 miles south of their release site. After several months, they burst onto the expansive open grasslands of the refuge to hunt elk aggregated in the thousands for all the world to see. Wolves attracted record numbers of visitors to the refuge that winter, and the usually quiet Refuge Road, which offers the only public passage inside refuge borders, suddenly experienced what might be called traffic.

Wolves were the catalyst for one of North America's more distinctive wildlife spectacles ever, but they were not the focus. On Valentine's Day, a group of wolf watchers glimpsed a mountain lion ascending Miller Butte on the National Elk Refuge, an unusual crescent-shaped outcrop of rock erupting at steep angles, just several miles northeast of the nearest neighborhood in "East" Jackson. At first, the group debated whether it really was a mountain lion, so short was the sighting, which only one among them had seen. Then the animal reappeared from a cave halfway up the hill in the company of three kittens.

Word of the mountain lion family spread like wildfire in a drought. Every morning and evening, a growing throng of people stood vigil on Refuge Road about 125 yards from the entrance to the cave in the hopes of glimpsing, photographing, and filming wild mountain lions. Refuge law enforcement erected signs reading "Parking for mountain lions" in order to alleviate the congestion. In the evenings, up to 500 people crowded together, packing snow at the base of Miller Butte.

"You could hear a pin drop, it was so quiet," Kevin Painter, among the refuge's law enforcement officers at the time, said of onlookers. "Everyone was so respectful." The mountain lions, for their part, were incredibly dependable, providing regular appearances before the cave, where they snuggled against one another and played. Kittens absconded with bits of hide from nearby kills, using them to play tug-of-war and cat-and-mouse for days on end. Over the course of forty-two days, an estimated 15,000 people witnessed wild mountain lions cuddle, lounge, play, nap, and eat wild prey.

Never before in the history of North America had a wild mountain lion been so accessible for so long, or been enjoyed and appreciated by such a crowd. The Miller Butte spectacle was absolutely transformational for those who were lucky enough to partake. Kevin Painter remembers bumping into people in town several years after the event, and if they were wildlife people, the subject of the Miller Butte lions would come up. "Were you there?" they would ask each other, groping for reconnection. When reunited with others who shared the experience, some broke down, glazed over, and devolved into stammering. Some shed tears before the Jackson post office, transfixed by joyous memories of standing in the cold with hundreds of others watching something as rare and wonderful as mountain lion kittens playing.

Direct experience simultaneously speaks to hearts and minds and is arguably the best means of changing people's belief systems and increasing tolerance for large carnivores. The transformational power of direct experience is also one argument for supporting wildlife tourism, which additionally provides economic incentives for local people who might otherwise persecute the same animals. Tourism places a value on wildlife for being wildlife, and thus has become a popular strategy embraced by conservation scientists and organizations in promoting predator

A modest crowd gathers at the National Elk Refuge in March 2018 during a rare mountain lion appearance. Two cats were visible about 500 yards from the road, where they fed on an elk carcass for several days. It was reminiscent of the famous 1999 event that occurred just down the road and attracted some 15,000 people over forty-two days. (Photograph courtesy of Ryan Dorgan / Jackson Hole News and Guide)

conservation. Around the globe, tourism for sharks, crocodiles, and big cats has blossomed into an annual industry that generates billions of dollars. In the Greater Yellowstone Ecosystem (GYE) surrounding Yellowstone National Park, for example, wolves bring in an estimated *additional* $22 million dollars every year to communities in Montana, Idaho, and Wyoming.[9] A single bobcat in Yellowstone, visible to winter tourists, was estimated to bring more than $300,000 annually to the same communities.[10] People travel to see big animals with big teeth more than any other kind of wildlife.

Opportunities for people to watch wild mountain lions are unfortunately few and far between. Tourism as being tested in Patagonia is unlikely to be compatible with chasing and hunting mountain lions in the USA. More importantly, Patagonia tourism appears reliant upon open habitat, which is less common in North America. There are a few houndsmen, however, who offer opportunities for photographers to join them for mountain lion safaris. In such instances, trained hounds chase mountain lions up trees so that they can be approached by paying clients. These opportunities appear equally transformational for their participants: "Nothing can compare to the rush of being up close and personal to a beautiful wild animal you have successfully tracked," reported one such participant who attended a "photo-hunt" with Bodhi Expeditions in Colorado.[11] Unfortunately, not everyone will be able to accompany a houndsman to see a wild mountain lion, so what can we use as surrogates for direct experience?

Photographs and film. People respond powerfully to narratives about the challenges that individual animals face, and when accompanied by compelling imagery, stories are proven conservation tools that educate, engage, and motivate people into action. There's a reason that Sir David Attenborough is listed among the world's most influential conservation heroes. He's not a scientist in the traditional sense, although his knowledge of the natural world far outstrips that of most academics. He's a television host of natural history programming and a filmmaker bringing the wonder and beauty of our planet to millions and millions of people around the globe every year. Steve Winter, Drew Rush, Sebastian Kennerknecht, Jeff Hogan, and Casey Anderson are all photographers and filmmakers who have shared incredible imagery of mountain lions and their secret behaviors. Natural history films produced by NatGeo Wild and the BBC, one of which was recently narrated by Sir David Attenborough, are teaching the masses about the hitherto-unknown lives of mountain lions. Films and photographs have real impact.

Biologists, too, benefit the species when they select research questions emphasizing that mountain lions are living, breathing individuals exhibiting distinctive motivations and behaviors. As the old adage

An adult female mountain lion and young kitten followed at the time by Panthera scientists, caught on a motion-triggered camera set in northwest Wyoming by professional photographer Steve Winter for the article "Ghost Cats" that appeared in the December 2013 issue of *National Geographic*. (Photograph courtesy of Steve Winter / *National Geographic*.)

teaches, we cannot love what we do not understand. The more we know about mountain lions, and the more we can share what we know about mountain lions, the more people seem to appreciate them. In an assessment of 3,657 articles written about mountain lions in the media from 2000 to 2015, there was an uptick in positive media about the species. More importantly, educational articles inclusive of researchers and research projects increased over the same time period, suggesting that education really does help.

Hunting may also play a role in supporting mountain lions, even if it fails to promote tolerance for large carnivores. Many biologists and

Are state actions increasing the risk of cougars attacking people?

by Rico Moore

The front cover of the September 12, 2019, issue of *Boulder Weekly*, a Colorado periodical, featuring an article by Rico Moore about the potential link between mountain lion hunting and negative human–lion encounters. It is representative of an uptick in positive media about mountain lions, and the importance of researchers providing information to keep media accurate. (Used with permission of *Boulder Weekly*.)

wildlife managers argue that permitting carnivore hunting is a symbolic gesture—it is an extended hand to people intolerant of carnivores, and those who might work to block or undermine carnivore conservation efforts. But there's no evidence that this works. In fact, it's a case of flawed logic because it assumes there is no distinction between people with negative attitudes about large carnivores like mountain lions and wolves, and those willing to flout the law to kill them illegally.

"Blood does not buy goodwill," concluded two scientists in a review of the effects of increasing legal predator hunting on illegal poaching activities and tolerance for large predators more generally.[12] Both initiating legal hunting of wolves in Wisconsin and liberalizing legal hunting of brown bears and wolves in Europe failed to reduce poaching activities or to lessen negative attitudes about either large predator. As a result, legal hunting actually increased mortality in these populations, contributing to their declines. Therefore, we cannot expect mountain lion hunting to change the beliefs or behaviors of people who would flout the law to kill mountain lions anyway, whether for revenge, pure hatred, or some other twisted set of values.

Haters aside, however, hunting does increase support for mountain lions, at least among houndsmen and -women. It's the age-old question as to whether hunters respect the animals they follow and fall in love with their quarry. It's a question many wildlife advocates find impossible to comprehend due to its irony. This is because some advocates refuse to acknowledge that there are different kinds of hunters.

In 1976, Stephen Kellert published a report called *Attitudes and Characteristics of Hunters and Anti-hunters and Related Policy Suggestions*, in which he categorized three types of hunters. He estimated that 44 percent of hunters were "utilitarian" and primarily hunted for food. Thirty-nine percent were "dominionistic," and primarily hunted as a means of participating in competition, demonstrating prowess with weapons, and exhibiting mastery over animals. The third and smallest group he called "naturalistic." Naturalistic hunters primarily hunted as a means of connecting with and participating in nature—these are the ones who confess to falling in love with their quarry.

Kellert was particularly struck by one pattern in his research. He believed that a person's view of hunting was "a kind of barometer for assessing people's much broader understanding of the natural world." Based on his data, hunters expressed a greater interest in interacting with wildlife than nonhunters, and expressed greater affection for wildlife and a lack of fear of wildlife than did nonhunters. In short, hunters had greater direct experience with nature than the typical nonhunter, and as a result, they had greater appreciation for the natural world.

In follow-up research conducted in the late 1970s, birdwatchers, wildlife advocates (which Kellert defined as different from animal welfare advocates), general scientists, and naturalistic hunters proved the most knowledgeable about wildlife. Backpackers and utilitarian hunters scored in the middle, and dominionistic hunters and anti-hunters ranked near the bottom.[13] Today, dominionistic hunters dominate not only their quarry but also media coverage representing hunting to the American people. As a result, dominionistic hunters driven by money and power wielded by the likes of Safari International, are increasingly damaging hunters' reputations and our natural resources as a whole. Dominionistic hunters are unlikely to be interested in the conservation of mountain lions or any wildlife, and, unfortunately, evidence suggests that dominionistic hunters are a growing proportion of US hunters.[14]

Mountain lion hunters likely reflect the diversity of hunters found in the larger hunting culture, but there are also categories of mountain lion hunters based on their methods. People hunt mountain lions on foot, with predator calls, and with traps and snares. But the method that stands out is hound-hunting, a centuries-old hunting tactic in which humans team with dogs to follow scent and visual clues to chase their quarry. Most houndsmen and houndswomen revel in working with dogs—even more than they do in hunting any particular type of prey. In *When the Dogs Bark "Treed,"* Elliott S. Barker writes: "There is nothing that gives one more pleasure and satisfying enjoyment than watching and listening to a pack of well-trained dogs as they work out an old, difficult trail." Houndsmen, and houndswomen too, have proven to be valuable advocates for mountain lions.

The conservation roles of hound-hunters on behalf of mountain lions is actually quite remarkable, and may be distinctive around the world. Even with all the hunting of wolves across the Northern Hemisphere, there isn't a modern-day, hunter-led advocacy for wolves. The sentiment is generally that the more wolves we kill, the better. Coyote hunters don't form groups to try to protect coyotes, nor do bear hunters advocate for bears. In contrast, houndsmen in both Montana and Wyoming are largely credited with killing legislation that would have allowed leg-hold traps and snares to be set for mountain lions. Houndsmen's associations in Montana and some other Western states advocated for ending the bounty system and switching to managed hunting seasons in the mid-twentieth century. And every year, houndsmen across the West pit themselves against deer and elk hunters, and request lower mountain lion quotas, in an effort to see mountain lion populations grow.

"Unlike other groups who would like to see large predators entirely wiped out or those who would like to see them completely left alone with no management whatsoever, the houndsmen of this state hope to suggest and help implement sound scientific and common-sense practices so that we may all enjoy Wyoming's wildlife resources," pronounced Tex Adams, president of the Wyoming Federation of Houndsmen, at the Wyoming Game Commission meeting discussed at the start of chapter 5. J. D. Downer, president of the Wyoming Houndsmen Association, followed Adams and added, "Everybody across the state is seeing a huge decline of lions. . . . We want to see more source areas [for mountain lions] . . . just lower quotas across the state." Both

A mountain lion cornered in a rock cleft by scent-trailing hounds. More typically in North America, mountain lions retreat to trees.

houndsmen's organizations advocated for reducing mountain lion hunting, even at the expense of losing hunting opportunities for themselves.

The unique behavior of houndsmen on the behalf of mountain lions may in part be explained by their unique hunting methods. Hound-hunters can chase and "catch" mountain lions, and then walk away without killing them. Therefore, they can reduce hunting quotas without giving up what they enjoy most—chasing and catching cats. Many experienced houndsmen, in fact, choose to stop killing mountain lions completely. Cal Ruark, former president of the Bitterroot Houndsmen Association in Montana, is among them. "A mountain lion to me is just an amazing animal. I love 'em. It's just something to see them in the tree, or in the rocks or wherever you get them bayed up. I don't care if it's a spotted cub or a 160-pound tom."[15]

It's for these reasons that Jim Williams, wildlife manager for the Montana Department of Fish, Wildlife & Parks, wrote in his book, *Path of the Puma*: "By allowing a certain amount of hunting, we guarantee a constituency who will speak for wild cats. What's best for the cat, ultimately, may be to hunt a few—in order to secure public support for the rest. That's not science. That's people and politics. And it's real."

The irony is not that hound-hunters might fall in love with their quarry, but that we need mountain lion hunting to protect mountain lions from the influence of ungulate hunters on state wildlife management. Like it or not, hound-hunters, who are typically conservative white men (and some woman as well), are politically more influential in participating in and guiding state wildlife agency legislation and regulation than are wildlife advocates or people from minority cultures, all of whom sit outside the normal purview of state wildlife agendas. This is by no means a sanctioning of the status quo—wildlife management is a broken system in need of attention, as we'll discuss in chapter 8. But it is a frank appraisal of the reality right now. Hound-hunting helps mountain lions because houndsmen and houndswomen help mountain lions.

Barry returned to empty his beetle cups one week after he set them in the ground, and then again the week after that, and the week after that, until he had accumulated nineteen weeks of sampling at eighteen carcass sites. Each time he emptied a cup, he refilled it with water and dish soap. Occasionally he had to rebuild his beetle traps after a bear had come through, overturning everything and scattering his equipment.

Day after day, Barry returned to the office with his precious catch, where he worked beneath a small pavilion erected in the yard so that the putrid stench of death that seeped from his collection jars no longer impacted his peers. Over the course of a summer and early fall, Barry collected and counted 24,209 adult beetles, of which 14,928 were large black-and-yellow northern carrion beetles. The remaining species were more difficult to categorize and required aid from experts. He sent his beetles to Dr. Mike Ivie at Montana State University, the world's expert on Yellowstone beetles. Mike passed the assignment along to Lisa Seelye, who, over several months of diligently peering into microscopes, helped Barry identify an astounding 215 unique beetle species caught in his cups.

Carrion, which is science talk for dead meat, supports a surprising abundance of life. Just as decomposing wood supports mycelia essential to forest health, so too does dead meat support vibrant life among both mobile and sessile flora and fauna. Carrion increases an ecosystem's health and its ability to heal itself after a catastrophic event. This ability to bounce back after flooding or wildfire or a disease outbreak is called *resilience*, just as the mountain lion superorganism bounces back after persecution.

With regard to carrion, however, it's useful to think of resilience in terms of linkages in food webs. Traditionally, food webs comprise big predators like mountain lions and wolves that eat large prey like deer and elk, and smaller predators like bobcats and foxes that eat smaller prey like songbirds, squirrels, and mice. The linkages in the traditional food web are few—predators connect to their prey and nothing more. But a food web with dead deer—that's another story. Panthera's Teton Cougar Project documented thirty-nine species of birds and mammals that feed on mountain lion kills in northwest Wyoming.[16] Add a dead deer, and suddenly mice, squirrels, chickadees, insects, and many other species are directly linked to deer, because they eat carrion. Mountain lions thus become indirectly linked to woodpeckers, Steller's jays, and long-tailed weasels because they provision them with food. In short, carrion exponentially increases the linkages in food webs and the pathways moving energy throughout natural systems. So if a wildfire or disease outbreak severs several pathways in an ecosystems food web, carrion supports numerous alternative avenues for energy to move around and heal the wound.

Not all carrion is equal, either. Big chunks of dead meat are more beneficial to ecosystems than small chunks, because they support more linkages and more varied linkages in food webs. For this reason, large predators provide distinctive and important services. Mountain lions kill prey up to eight times larger than themselves, and then keep the carcass intact as they feed. Wolves, by contrast, generally kill more prey than mountain lions per unit area, but they dismantle their prey into smaller chunks, reducing the ecological benefits of larger carcasses. Mountain lions, it ends up, are the veritable McDonald's of the animal kingdom. They, too, feed the masses.

Every day, mountain lions provide roughly sixteen pounds of meat per forty square miles (100 km^2) to their ecological communities. Across their 8.8-million-square-mile range across North and South America, mountain lions contribute roughly 3.3 million pounds (1.5 million kilograms) of meat to other animals every day[17]—that's more than 1.2 billion pounds of meat each year. McDonald's restaurants in the United

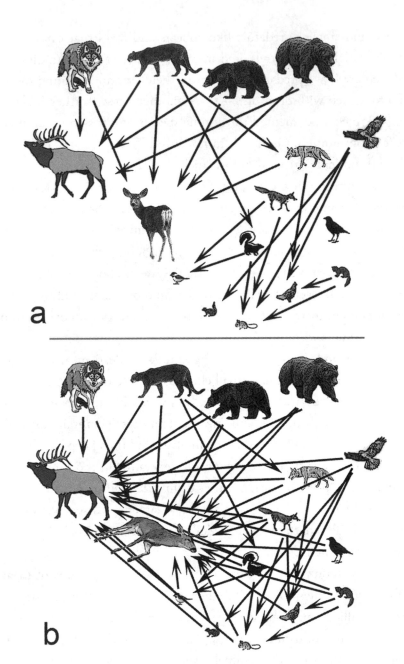

a

b

(a) A simplified conceptual food web with predators eating their respective prey. (b) The incredible increase in food-web linkages when we consider dead ungulates in the same system.

A black bear in northwest Wyoming pauses to stare at hovering ravens unhappy to have their food commandeered by another. Both species were benefiting from an elk killed by a mountain lion but stolen by the bear. Panthera scientists documented thirty-nine species of birds and mammals feeding on mountain lion kills in Wyoming.

States, with their billions and billions of hamburgers served over the last half century, only distribute about one billion pounds of red meat each year.[18] And whereas McDonald's beef-buying practices threaten forest health and biodiversity, mountain lions fortify forests and bolster biodiversity. Nutrients from the carcasses of prey killed by mountain lions soak into the earth and are sucked up by surrounding vegetation, elevating crude protein and other minerals that fortify plants and the animals that eat them.

For beetle communities, mountain lion prey proved to be entire neighborhoods and cities, not just supermarkets. Beetles were communing and mating, seeking protection from predators, hunting other invertebrates, and raising their larval young, all within the confines of dead animals provided by mountain lions. For these reasons, Barry and his team were able to argue that mountain lions were *ecosystem engineers*—species that create or modify habitat for other animals.[19]

American beavers are the classic ecosystem engineer. A stream without beavers tends to be uniform in water depth, speed, and temperature. Throw in some beavers, and they begin to build dams and generally make a mess of things. As a result, water slows down and becomes deeper and warmer where the stream is clogged by beaver dams to form ponds. Deep, still beaver ponds soak up sunlight and support different fish, plant, and bird life than the cooler, fast-running portions of the same stream. Streams with beavers have more kinds of habitat and therefore support a greater diversity of flora and fauna—collectively, this is what we call biodiversity.

The catch phrase *keystone species* is generally well known, even if the concept remains slightly murky. The keystone in an arch is the arch's linchpin, providing structural strength and integrity to the entire arch— remove it, and the structure topples and falls. Ecosystems function in much the same way. Keystone species are animals that disproportionately influence their environment and are often those that best support ecosystem health and biodiversity. Remove them and systems suffer.

Not every keystone species is an ecosystem engineer, but every ecosystem engineer is a keystone species. Mountain lions disproportionately hold ecosystems together. They influence the habitat selection of their prey, and almost certainly cause trophic cascades.[20] Mountain lions create more large carcasses than other predators. They feed more mammals and birds than do any other predators, increasing the number of animal interactions that are so essential to maintaining ecosystem resilience. And on top of this they are ecosystem engineers as well. Mountain lions deserve conservation attention to ensure the sustainability of healthy ecosystems upon which we depend.

Given how little has been researched regarding the positive contributions made by mountain lions to their ecological communities, it's startling how much they do, especially in comparison to other top predators. To quote veteran mountain lion researcher John Laundré, "Cougars are not minor actors, not stand-ins, not faces in the crowd. Their role is significant, and many feel that if not the stars of the show, cougars are at least major supporting actors."[21] And this is good news for areas seeing the recovery of mountain lion populations, like the Midwestern United States and the grasslands of Patagonia. Mountain lions may complicate the political landscapes they inhabit, but they also benefit the biological communities in which they play an important part.

Social scientists have yet to determine the definitive way to build tolerance for large carnivores or to change people's values and belief systems. And perhaps they never will, given that people are complex, varied animals that act unpredictably. Nevertheless, we should abandon the old approach of listing statistics about how people are more likely to die by thousands of other means than by mountain lions as a way to mitigate fear. Nothing suggests that such an approach works, and indeed it may be counterproductive. Instead, we need to invest in ecosystems services research emphasizing the economic value of mountain lions, as well as additional research emphasizing the importance of mountain lions to healthy ecosystems and human communities. Simultaneously, we need to invest in media projects and education campaigns to spread the results of this research and to highlight the intrinsic value of individual mountain lions to everyday people.

The question then becomes: Who should be responsible for caretaking mountain lion knowledge? Should we trust it to state wildlife managers, who are public agents but do not prioritize mountain lions . . . or to wildlife advocates, most of whom have never seen a wild mountain lion . . . or to mountain lion hunters, most of whom have killed at least one lion? Or should the responsibility be shared across stakeholders? Who should be responsible for teaching people about mountain lions, and ultimately who should decide their fate?

In terms of education, it would seem prudent for mountain lion biologists to shed their traditional roles as aloof vessels of knowledge and to participate in the dissemination of accurate information to counter "alternative facts" and oppositional pundits. Media coverage on the Florida panther, for example, proved more positive and more accurate when it included press releases and quotations of managers and biologists who worked with panthers.[22] Biologists in India go even further and provide media training and education events on leopards, which has increased not only the accuracy of the information presented about leopards in media, but also the positive portrayal of leopards in articles.[23]

The theme of the next chapter is *who* participates in mountain lion management, and who doesn't. State agencies manage wildlife for the public trust, but in practice, a great many people feel excluded from their decision-making processes. One critic described current management as overly "expert-based, control-oriented, and instrumentalist,"[24] and another as one of "the remaining bastions of bureaucratic, customer-oriented, scientific management."[25] Mountain lion advocates and houndsmen, for example, are clear stakeholders who want to participate in mountain lion management, yet both feel largely ignored by state agencies. Fostering greater inclusivity and building bridges among stakeholders—clearly, these are two parts of the solution to the frustration and anger born of the cougar conundrum.

CHAPTER 8

The Money behind Mountain Lion Management

"It's fair to say . . . lion management is consistently the most contentious management issue that we deal with. . . . Some houndsmen, other lion advocates, preservationists, don't honestly trust that the department either can or is willing to conserve lions, to protect lions, while also harvesting lions. Some of that distrust has been fairly well earned."[1] So it was that Jay Kolbe, wildlife biologist for Montana Fish, Wildlife & Parks (FWP), addressed an uneasy audience during the public meeting discussing proposed changes to mountain lion hunting regulations held on March 13, 2014, in Three Forks, Montana. Mountain lion management in Montana had always been a prickly subject, but several new bombshells fell in 2013 and 2014, igniting yet another local cougar conundrum.

The first was new research led by FWP biologist Kelly Proffitt estimating mountain lion abundance in the Bitterroot Valley. It was a new method based on genetic samples gathered from hair and scats collected while following footprints in snow, or from muscle tissue from animals chased up trees by houndsmen and shot with biodarts that collect a tiny sample.[2] The goal of the research was to test a potential avenue for determining mountain lion abundance without having to catch them and fit them with collars and invest hundreds of thousands of dollars in following them, as FWP had done before. Proffitt's estimate was twice that of estimates published just several years earlier by veteran lion biologist, Hugh Robinson, using the traditional method of collars,[3] and who along with several others were also coauthors on Proffitt's new research paper. Part of the difference in mountain lion numbers was due to the fact that the genetic method included resident mountain lions that lived in the area *and* transient youngsters passing through in search of a territory to call their own, which many argue is a better approach to estimating overall lion abundance. But it didn't matter that the two methods were like comparing apples to oranges, they were ammunition in the ongoing debates about mountain lion hunting quotas. Those that were anti-mountain lion rallied around the science led by Kelly Proffitt, while those who were pro-mountain lion clung to the science led by Hugh Robinson.

The second bomb fell fast on the heels of the first. Dan Eaker led new research reporting that mountain lions were the primary predator of elk calves in Bitterroot Valley where Proffitt had conducted her study, and where elk populations had been struggling in recent years.[4] Worse, even though Eaker's research suggested that calf survival was primarily driven by forage and precipitation, it also provided some evidence that calf survival was impacted by predators in winter. This pronouncement opened the door for local managers to contemplate whether controlling predators might be a means of increasing elk calf survival, and ultimately increasing elk numbers. To add insult to injury, Eaker's research cited Proffitt's study to highlight that the local mountain lion populations might be much higher than anyone had suspected.

The response from elk hunters was immediate. Waving Proffitt's and Eaker's research like banners in war, they demanded mountain lion blood. FWP responded just as quickly, devising a new mountain lion harvest program meant to knock back the mountain lion population by 30 percent over three years, in an attempt to aid local elk. And that's when the houndsmen really got angry.

"What we got was keyboard cougars and paper pumas,"[5] claimed veteran houndsman, Rod Bullis, referring to the Proffitt research. Houndsmen claimed the new mountain lion population estimate was completely misinformed and shouldn't be used for any management decisions. Now armed with Proffitt's research, elk hunters pushed back, saying that the mountain lion population had grown out of control. Accusations were leveled and tempers flared. FWP found themselves in a quandary and were unable to move forward with a mountain lion management plan that both elk and mountain lion hunters would support.

Trail camera images providing a comparison of the proportions of larger adult male and smaller adult female mountain lions, the difference that drives hunter preference for males. (Original photographs courtesy of Pete Alexander / Craighead Beringia South-Panthera.)

It is important to note that mountain lion research is rarely done to provide answers for a specific management decision. For this reason, people pull research from its original context and integrate it with their personal values to interpret the results in a way that justifies their point of view.[6] It is human nature to seek evidence to support one's own beliefs and to ignore evidence that might threaten one's convictions. And today, the Internet and social media provide a plethora of science to pull from to support just about any perspective one chooses. Scientists are as prone to mistakes, quality control problems, and hidden agendas as the humans who interpret the research. "Predatory" online journals have sprung up like weeds, and for a price, they are willing to publish anything anyone has written and claims to be science. There is now more opposing research published than ever, especially given that the average person cannot differentiate reputable scientific journals from counterfeits.

On social media, it's become a common occurrence to witness hunters and advocates flinging "science" at each other like stones. In the case of the Bitterroot Valley, the scenario was even worse—the opposing teams were each wielding different, but defensible, science led by the very same organization, the Montana Department of Fish, Wildlife & Parks. Nevertheless FWP had one more card to play. They were going to try something unprecedented, at least in mountain lion management. They invited local people to participate in a workshop during which they would recommend mountain lion quotas to FWP for Region 2, rather than the other way around. It was a remarkable opportunity for public participation in what typically operates as a closed process.

The ferocious anger and frustration that has become so characteristic of the cougar conundrum isn't caused by the recent increase in mountain lions living in the United States and Canada, nor even by the fact that we hunt them. The root issue is the growing division between government bureaucrats and everyday people. It's the concern that Jay Kolbe tried to address at the meeting described at the opening of this chapter. Many people, houndsmen and mountain lion advocates among them, believe that state wildlife agencies ignore their concerns about overhunting mountain lions and exclude them from decision-making. As a consequence, mountain lions suffer because of it. A key component to resolving the cougar conundrum is increasing inclusivity in decision-making. Who, then, would FWP accept into their working group?

As defined by wildlife professionals today, *management* is not equivalent to *conservation*, although the two disciplines overlap. The distinction is a century old, when early wildlife managers and biologists began to distinguish between conservation and preservation. *Conservation*, as coined by the first chief of the United States Forest Service, Gifford Pinchot, and environmental hero Aldo Leopold, meant the sustainable "wise use" of natural resources within a management framework. *Preservation*, in contrast, emphasized nonuse and the complete protection of resources and ecosystems.

Today these definitions are different, contributing to the confusion over what conservation means. Once in a while, an agency staffer might use *preservation* as defined long ago, as did Jay Kolbe during the opening of this chapter, but largely the word has fallen away. Today, people say "wise use," "sustainable use," or "*applied* biology or ecology" to refer to natural resource management. Wildlife professionals now use *conservation* to emphasize the protection of ecosystems and biodiversity more broadly. Thus, what was "conservation" a century ago is now sustainable use, and what was preservation a century ago is now more akin to modern conservation.

The division between "wise use" management and modern-day conservation became more apparent when the location of conservation science innovation and leadership moved from within state and federal agencies to universities and other external organizations in the early twentieth century. This transition allowed the scientific basis for conservation to propagate outside the bureaucratic constraints sometimes imposed by wildlife agencies under the influence of politics and special interest groups. It also set forth parallel trajectories in science. One focused on wildlife management and emphasized the sustainable, wise use of natural resources, whereas the other emphasized conservation innovation focused on understanding, preserving, and restoring biodiversity as a means of ensuring ecosystem health and resilience. Today, these two distinct disciplines of science sometimes find themselves at odds with one another.

Modern wildlife management is primarily a human enterprise rather than a scientific one. In its simplest form, it is the decision-making process guiding wildlife populations toward some desirable goal. These decisions include any knowledge that science might bring to bear on the subject, but they are heavily influenced by public values, the interests of state wildlife constituents (including official and unofficial lobbies), and, with regard to mountain lions, objectives set for other species such as deer and elk. Modern wildlife management also follows a business model for hunting clientele, as proposed by George Bird Grinnell and Teddy Roosevelt, and therefore caters to customer satisfaction.[7] Add to this complexity the fact that mountain lion management is also always made in the face of imperfect knowledge about the effects of management decisions on wildlife populations. Sometimes mountain lion management includes a little science and a lot of public value; on other occasions, science holds greater sway.

Let's consider the Bitterroot Valley cougar conundrum fueled by Proffitt's and Eaker's research as an example. How are elk calves doing in that part of the world? Evidence would suggest that they're doing just fine. For the first three months on the ground, elk calf survival was higher in the Bitterroot study system than the average calf survival across twelve Western populations, and identical to other populations with an equal diversity of predators across the full length of summer.[8] Local calf survival in winter was high as well, with an average of 73 percent of them surviving until spring. The fact that mountain lions were the most common predator of elk calves was not surprising, given that coyotes and black bears only eat elk calves for a very short window after they are born, and that Eaker and his science team reported that mountain lions were two to five times more abundant than wolves in the system.[9] Based on these facts, one might question how FWP determined that they needed to implement a 30 percent reduction in mountain lions in the region. The answer is the very real impact of people and politics on wildlife management decisions, including, in this case, pressure from the Ravalli County Fish & Wildlife Association and the larger elk hunting community to improve hunting opportunities.

On March 19, 2014, FWP sent a news release to local media outlets in the hunt management area designated as Region 2, which included the Bitterroot Valley. It was an invitation to people with a "demonstrated interest in mountain lion management" to participate in a four-day workshop, during which the working group would come to a consensus on regional hunting quotas for mountain lions the following year.[10] More specifically, FWP "sought official representation" from the Ravalli County Fish & Wildlife Association, which represented elk interests, and the Bitterroot Houndsmen's Association, which represented the opposition in the fight over mountain lion quotas.

FWP received seventeen applications from which they selected twelve participants based upon the following criteria: they wanted two people each from the four major watersheds in the region, and they wanted "stakeholders in wildlife management in Region 2." Participants were also required to commit to four full days of workshop together, and to agree to entering the workshop with the goal of reaching a consensus for mountain lion hunting quotas among the group.

FWP did not accept the applications they received from mountain lion biologists from other areas, representatives of wildlife advocacy organizations, or others who did not have local knowledge of the issues. Ultimately, hunters were prioritized over any other constituent group, and those hunters who held sway with their counterparts were prioritized most to increase buy-in from the local community in the process. Other members of the public were welcome to attend the workshops and observe, as well as to make comments during short periods devoted to public input that were allotted each day.

Dr. Mike Mitchell of the Montana Cooperative facilitated the mountain lion working group, as they were named. He led participants through a process called structured decision making (SDM), during which personal values and scientific knowledge were disentangled so that participants could more clearly assess the influence and relevance of each on the decision they needed to make. SDM begins with values, as values are generally the foundation of both conflict and decision-making.

On day 1 of the workshop, the group began by drafting a problem statement, composed of everyone's interests and concerns. Although the group battled to find common ground, they eventually came up with a statement highlighting the following key elements:

First, FWP is responsible for managing healthy, sustainable wildlife populations.

Second, FWP is responsible for supporting the continuation of Montana's hunting heritage.

Third, there is disagreement among stakeholders with regard to (1) the number of mountain lions on the landscape, (2) the desired number of mountain lions, and the male–female composition of the desired

mountain lion population, and (3) the impacts of mountain lion predation on ungulate populations.

Fourth, there is insufficient information available about living with
mountain lions.[11]

These details of the working group may seem unnecessary, but they
are instructive in understanding the importance of inclusivity on collaborative outcomes. (Keep this in mind, as we'll revisit this idea at the end
of the section.) Based upon their problem statement, the group created
the following shared objectives for any mountain lion hunting program
they devised over the course of the workshop:

1. Maximize satisfaction of resident lion hunters.
2. Maximize satisfaction of nonresident lion hunters.
3. Improve ungulate numbers in at-risk districts in Region 2.
4. Maintain acceptable densities of mountain lions for
    a. ungulate hunters
    b. landowners
    c. houndsmen
    d. outfitters
    e. nonhunters
    f. nonresidents
    g. urban–wildlife interface
5. Improve sportsman support for mountain lion hunting.
6. Improve public support for mountain lion hunting.

Once the group had articulated the problem tasked to them and had
created shared objectives that were inclusive of the values of all people
participating in the workshop, they set about creating five alternative
mountain lion harvest strategies for the twenty-seven hunting districts
in Region 2. The first they called "Status Quo," and it followed the
current FWP plan to continue to reduce the mountain lion population
for the third year in a row. The second aimed to maintain the mountain

lion population at its current levels. The goal of the third was to reduce mountain lion populations where ungulates were of greatest concern. The fourth aimed to increase the mountain lion population, and the fifth, to increase the number of trophy male mountain lions.

By now the group was well into its second day of sweating and arguing, with absolutely no truce in sight. Their final task for day 2 of the workshop, before they could escape to vent their frustrations, was to rank their satisfaction with each of the five alternatives they had created on a 1–5 scale. For example, we would expect elk hunters interested in increasing elk numbers to be more satisfied with alternatives that reduced mountain lions than those that increased them.

At this stage of the workshop, a science team stepped in to assist the working group with visualizing the alternative mountain lion strategies they had proposed, in terms of defining how many male and female mountain lions would need to be killed to meet their objectives. Specifically, the science team created mathematical models to simulate the effects of different harvest strategies on the future of the regional mountain lion population. The point of these models was to help the group better understand the consequences of each of their five alternative management strategies for local mountain lions over time. At no point did the science team impose their own strategy. They played a support role, nothing more.

As the most contentious issue among workshop participants was which of the current mountain lion abundances (Proffitt versus Robinson) was accurate, the science team modeled the impacts of the five harvest regimes created by the working group on four hypothetical mountain lion populations: (1) a mountain lion population with the density reported by Robinson's research that focused on the number of resident adult animals, (2) one with the density reported by houndsmen in testimony to the Fish and Wildlife Commission, the governor-appointed individuals responsible for setting state wildlife regulations in April

An adult female in a tree in the Pacific Northwest, where she was captured with the help of hounds and houndsman Greg Jones as part of research efforts.

2014, (3) one with the conservative density estimate reported by Proffitt's research team, and (4) one with the median density reported by Proffitt's team. This way, everyone's beliefs about how many mountain lions currently existed in the region were included.

After the second day of workshops, the group went on a two-week hiatus to give the science team time to develop the appropriate models to match the alternative proposals created by the working group, and then to forecast their impacts on the four hypothetical mountain lion populations over the next five years. On day 3 of the workshop, the science team presented their results. The presentation proved to be a lynchpin moment, as it transitioned the group's focus from a fight over how many mountain lions were on the landscape to one of the impact of the proposed management strategies on the local mountain lion population.

First, the presentation illustrated that, regardless of what mountain lion density a person believed was true, the impact of the five management alternatives was the same. In other words, it didn't matter how many cats were on the landscape; heavy hunting of female mountain lions always sent the population on a downward trajectory, whereas reducing mountain lion hunting or more selectively hunting males over females always increased the population just slightly. Second, the future projections presented by the science team demonstrated that the effects of the five different harvest strategies proposed by the group were not so different from each other as participants expected them to be—thus the alternatives "maintain the mountain lion population" and "increase trophy males," for example, were in fact very similar in how hunting would be managed on the ground.

"It showed the value of bringing rigorous science to bear on the problem," explained Mitchell, when reflecting on the group's evolution. "Densities weren't the problem, different quotas weren't the problem. It was how people felt about them. And so, if you talked about the quota that was intended to reduce the lion population, then the lion hunters would get upset. But when they saw exactly what that quota did to the lion population, it kind of took the fear factor out of it. Then it was just a matter of finding a compromise solution everyone could live with based on their values instead of biology."

Following the presentation, participants reviewed their own satisfaction with the outcomes of the five alternative mountain lion management strategies that they reported on day 2 of the workshop. Alternative 3, "reduce mountain lions in areas where ungulate populations are below desired levels," represented the best middle ground among opposing views in the group. Nevertheless, the simulations presented by the science team showed that alternative 3 would reduce the mountain lion population over time. This outcome would serve the goals of elk hunters, but ultimately fail to meet the desired outcomes of mountain lion hunters. Thus, the group agreed to modify their second alternative, "maintain the mountain lion population," by redistributing the hunting quota in ways that both satisfied elk hunters and maintained the mountain lion population at the regional scale.

From there, participants engaged in focused horse-trading. They knew how many lions needed to be harvested to reach their agreed-upon objective of aiding elk while simultaneously trying to maintain the mountain lion population. They also knew the areas where the elk population were of greatest concern. So the overall hunting quota was negotiated and distributed in ways that satisfied both elk hunters and hound-hunters, by having heavier and more female-dominated hunting in elk areas, and lighter mountain lion hunting with less females in other areas to better maintain the mountain lion population in the region. Following the conclusion of the workshop, the group's final mountain lion management strategy was accepted by FWP and implemented for Region 2.

It was an amazing example of a state wildlife agency reaching out to its constituents to include them in the decision-making process, as well as how SDM helps people find resolution to what at first appears to be insurmountable obstacles. It also remains a shining beacon of hope for anyone frustrated with mountain lion management and conservation in our country today. Tools exist to bring people together and encourage collaboration over division. In cases where stakeholders are completely unwilling to work toward any compromise, there are other tools as well. Conflict resolution, mitigation by a third party, and joint fact-finding are all means of building bridges across chasms among opposing ideologies, values, and desired outcomes for the mountain lions with which we share the world. State agencies just need to employ these available tools more often, and be more inclusive in their invitations to individuals who might participate. Recall the importance of the minutiae of this case study at the start of this section. The details of the problem statement and objectives created by the FWP mountain lion working group, for example, reflected the people at the table, and an emphasis on "sustainable management" over conservation. Would these details have been different if the group had included more women, or a wildlife advocate, or a high school student? Of course they would. People ultimately decide outcomes, and thus the people included in decision-making are critical to the process.

While it certainly made sense for FWP to prioritize elk and lion hunters from Region 2 in their mountain lion working group to facilitate resolution and the passage of mountain lion quotas in time for the next hunting season, one should question their decision not to include at least one representative from the community of people who advocate for wildlife for tourism and wildlife watching, or who just appreciate animals for their intrinsic value. Today these types of people are often referred to as the "nonconsumptive" public, and they are the vast majority of Americans today. Whereas hunters consume animals because they remove resources from the system, wildlife watchers use the same resources without using them up.

In theory, state wildlife agencies serve every resident of their respective states, safeguarding natural resources held in public trust for all people. In reality, agencies are entrenched institutions that continually wrestle with broadening their mission statement to be more inclusive of more diverse perspectives while simultaneously maintaining good relations with the special interest groups that fund them. Some agencies have gone so far as to change their names—removing a word with hunting connotations like *game* and replacing it with *wildlife* to indicate a shift from game management to broader biodiversity conservation, but these symbolic acts do not always reflect rapid internal evolution.

The question of whom state agencies recognize as their constituents, and whom they do not, comes down to money and power. Revenues from the sale of hunting and fishing licenses in conjunction with federal subsidies gathered from the sale of hunting and fishing equipment form the vast majority of state wildlife-program budgets. Hunters and anglers disproportionately fund state wildlife programs, inclusive of the broad array of state wildlife activities ranging from traditional big-game management to broader conservation programs that support biodiversity and ecological restoration; therefore agencies tend to recognize them as their primary constituents and customers, and often cater to their wants and needs.[12]

Hunters are, and should be, proud of their support of state wildlife programs. Hunters also support conservation through other avenues. Ducks Unlimited, a hunting organization that prioritizes habitat protection, for example, has protected more than 14 million acres of wetlands and other duck habitat across North America over the last eighty years. Through lobbying, Ducks Unlimited has contributed to the protection of nearly 200 million acres over the same time frame. Nevertheless, there are two caveats relevant to our discussion about whom state agencies include in mountain lion management versus those whom they *should* include.

First, while hunters disproportionately fund conservation administered by state wildlife agencies, they do not disproportionately fund conservation activities more broadly. In 2016, people invested in local and national economies by spending $26.2 billion on hunting activities, $46.1 billion on fishing, and $75.9 billion while "closely observing, feeding, or photographing wildlife."[13] Nonhunters also invest in conservation through other means, including nearly $13 billion annually to environmental and animal nonprofit organizations and through support of academic conservation efforts that, in total, far outstrip the inputs of hunters.[14] In addition, nonhunters disproportionately fund the federal agencies responsible for maintaining wildlife and other natural resources for the public trust, including the US Fish and Wildlife Service, the US Forest Service, and the National Park Service, just as a function of their tax contributions and their overwhelming numbers when compared with the hunting public. The number of people in the US who hunt is declining. Today slightly less than 5 percent of American adults hunt animals.[15] As of 2016, there were an estimated 11.5 million hunters versus 86 million wildlife watchers in the USA.[16]

Second, hunters disproportionately fund state wildlife activities by design, as part of a system in which those currently in power exclude nonhunters for fear of losing power over decision-making surrounding wildlife resources. Excluding nonhunters was unlikely the intent when the current business model for funding wildlife management in the United States was created, but nonetheless it is a consequence. Historians suggest that the movement to secure funds for wise-use management led by wealthy, politically connected hunters and politicians began as early as 1919, as a continuation of the Teddy Roosevelt era of conservation. In that year, John Burnham, then president of the American Game Association, wrote, "If the young men of the next generation are to enjoy from the country's wildlife anything like the benefits derived by present outdoor men, we must be the ones that shoulder the burden and see that our thoughtlessness or selfishness does not allow us to squander that which we hold in trust."[17] A working group of interested parties explored several alternatives, including the creation of a new hunting stamp. Ultimately, they decided that the best path forward was to try to re-allocate an existing tax on firearms and ammunitions to wildlife conservation. After a few bumps in the road, powerful sportsmen found allies in two Democrats, Nevada senator Key Pittman and

Opposite: Public participation in hunting and wildlife watching in the United States in 2016. Percentages reflect the percentage of the adult population that reported engaging in these activities for different regions of the country. (Originally published in US Department of the Interior et al., *2016 National Survey of Fishing, Hunting, and Wildlife-Associated Recreation*, 29 and 41.)

Hunting Participation
(National participation rate: 4%)

Mountain 5%

West North Central 8%

East North Central 7%

New England 2%

Pacific 2%

Middle Atlantic 3%

South Atlantic 3%

East South Central 8%

West South Central 5%

Wildlife-watching Participation
(National participation rate: 32%)

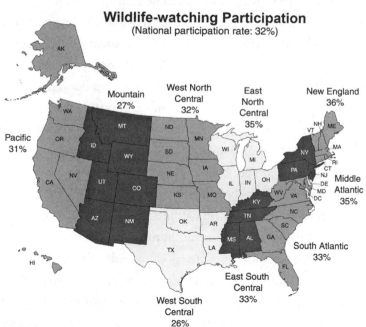

Mountain 27%

West North Central 32%

East North Central 35%

New England 36%

Pacific 31%

Middle Atlantic 35%

South Atlantic 33%

East South Central 33%

West South Central 26%

Virginia congressman Absalom Willis Robertson, who helped craft and present the final bill, called the "Federal Aid in Wildlife Restoration Act of 1937." Today, it is now more often referred to the Pittman-Roberts Act, or the PR Act, for its original sponsors.

The bill, and its modifications since, was ingenious in its scope and its written form, and remains one of the most powerful and influential pieces of conservation legislation ever written anywhere in the world (along with other American legislation such as the Endangered Species Act, the Clean Water Act, and the Clean Air Act, among others). It secured an existing excise tax collected by firearms retailers for the US government, and funneled it out of the general tax revenue overseen by the Federal Treasury into accounts administered by the US Fish and Wildlife Service instead. As the Pittman-Roberts Act stands today, it secures an 11 percent excise tax on firearms and ammunition, a 10 percent tax on handguns, and an 11 percent tax on archery equipment. More importantly, the PR Act secures these funds permanently so that they cannot be returned to the general treasury if, for example, a certain amount of time has passed and they haven't been spent. This component of the act, in particular, is something few similar appropriations achieve.

Annually, PR funds contribute $300–400 million to support, enhance, manage, purchase, conserve, and restore habitats and populations of birds and mammals, regardless of whether they are game species. On average, PR funds compose about 60 percent of state agency budgets. These funds are administered to states in different ways, but they include several clever components that further their conservation impact. To receive PR funds, state wildlife agencies must agree that fees collected from the sales of angling and hunting licenses cannot be spent for anything but conservation purposes—not road maintenance, prisons construction, state politician salaries, or any of the thousands of other ways public money might be spent. Second, it requires state wildlife agencies to match PR funds they receive by at least 25 percent, enlarging the pool of funds available for conservation significantly. Nowhere else in the world has such a system, and its creation and maintenance is something every American should celebrate and be proud of.

The Pittman-Roberts tax on firearms, archery equipment, and ammunition is the primary justification for the argument that "conservation has occurred on the backs of hunters," and that "American hunters are the greatest conservationists the world has ever seen."[18] It is also the premise of the "user-pay/user-benefit" system that justifies the special treatment awarded hunters and anglers by state and federal wildlife agencies.[19] The facts are plain—clever, forward-thinking hunter-conservationists in the late nineteenth and early twentieth centuries played a critical role in rebuilding North American wildlife populations following the widespread destruction of our natural resources by European settlers and the American hunters that preceded them. Nevertheless, it is misleading propaganda that hunters consciously and voluntarily invested in conservation through taxes funneled into PR funds. In fact, it wasn't hunting and hunters, but the *management* of hunters and hunting, that saved American wildlife.[20]

Hunters did not self-impose a new tax on firearms and ammunition to support conservation efforts, regardless of what the pundits circulate on social media. A small group of politically savvy hunter-politicians redirected an existing tax that was created to aid national debt. This means that hunters who bought a rifle and ammunition on the day before the PR Act was enacted paid the exact same amount as the hunters who bought a rifle and ammunition the day after. A portion of firearms sales goes to the manufacturers of the rifle and ammunition. A portion goes to the proprietor of the establishment that sells the merchandise. A portion is syphoned off by state taxes, and the only difference between purchases that preceded the PR Act versus those that followed it is in where the federal excise tax was directed. In the same vein, it is equally ludicrous to assume that everyone who buys bullets and semiautomatic weapons wishes to conserve wildlife. Only half the estimated five mil-lion members of the National Rifle Association (NRA) claim to hunt, for example.[21]

Given the incredible impact of the PR Act, however, nonhunters equally invested in conservation began to explore whether they could create a parallel funding stream for state wildlife agencies gathered from among everyone else spending time outdoors and benefiting from our natural resources. In 1976, Missouri passed legislation increasing their state sales tax by one-eighth of one penny for the state wildlife coffers. In 1996, Arkansas achieved something equally modest, and lesser victories have been seen in several other states, including Virginia, Minnesota, and Iowa.[22] The most concerted effort to create a parallel funding source to PR funds was the "Teaming with Wildlife" initiative proposed in 1996. It proposed a new 5 percent excise tax on all outdoor gear, to be secured for conservation efforts.

"The intent is to raise money for State conservation programs for nongame species like butterflies, songbirds, and turtles by broaden-ing the funding base to include those individuals who do not hunt or fish, but otherwise enjoy the outdoors. This would include millions of Americans who are bird-watchers, campers, hikers, and nature photog-raphers," explained Don Young, US congressman from Alaska, during the oversight hearing on the proposed legislation on June 6, 1996.

Teaming with Wildlife ultimately failed, even with support from some 900 organizations spanning those that might be described as land-scape conservation organizations, wildlife advocacy organizations, and sportsmen's associations. The organizations most opposed to the Team-ing with Wildlife proposal were the manufacturers of the products to be taxed, because it would increase costs for consumers, and therefore might impact sales and profit margins.

"Here are some important facts," explained Charles McIlwaine, vice president of the Coleman Company, maker of the ubiquitous stoves and lanterns found in nearly every campsite in every campground, and then vice chairman of the American Recreation Coalition. "A relatively low percentage of campers, hikers, and other recreationists report that they watch wildlife or go bird-watching. Recreationists . . . are generally satisfied with their present activities . . . and they do not appear to be very motivated to pay additional taxes for additional services."[23] No evidence was provided to substantiate these statements.

"Now, I don't have a statistical study to back this up," joked Thomas Dufficy, executive vice president of the National Association of Photographic Manufacturers, "but it seems to me when it comes to 'wildlife'—I put that in quotes—there are many more wildlife photos taken at bachelor parties than of bighorn sheep."

No one, it ends up, wants new taxes, and the failure to pass Teaming with Wildlife is a chilling reminder of what could have been the fate of the PR Act if it had been a proposal to create a new tax on firearms to aid conservation, rather than an appropriation of an existing tax. The statement made by the National Rifle Association at the Teaming with Wildlife oversight hearing, however, was the most troubling, as it made clear how politics and power were becoming obstacles to supporting wildlife agencies doing the work they were entrusted to do.

"The NRA must acknowledge and protect the vital role that the sporting community occupies in the conservation of this nation's fish and wildlife resources. . . . We cannot support such an initiative if it diminishes, transfers, or dilutes the sporting community's contributions to fish and wildlife conservation through PR and state licensing fees."[24] Translation—if Teaming with Wildlife funds supported state agencies, hunters would have to share their role in supporting state wildlife initiatives. In other words, diversifying funding for state agencies would threaten the current position of power held by hunters, gun manufacturers, and sportsmen's associations in influencing state wildlife activities on the ground.

Increasingly, the position expressed by the NRA during the Teaming with Wildlife oversight hearing is shared by the entire wildlife management machine, meaning the hunters and anglers themselves, the biologists and other staff that work for agencies, the professional societies that represent management-minded biologists and their interests, and their associated political connections with sportsmen's associations, hunter advocacy groups, and the NRA and other gun-rights organizations. These political alliances are so strong that they block and undermine efforts to diversify funding for state wildlife agencies or to increase inclusivity in wildlife management decisions. Further, those in power funnel funds to support research that promotes and justifies their management strategies over support for more-diverse ecological studies of mountain lions. State wildlife agencies also occasionally show bias in permitting research. The Wyoming Game and Fish Department recently returned a mountain lion research proposal, requiring the applicant to remove any mention of "conservation." In 2018, Idaho stopped accepting applications to study mountain lions and announced they will lead all such research themselves. As a result of these strategies, there are a never-ending number of studies of mountain lion predation on deer and elk that feeds the propaganda machine of current management. In contrast, one can count the number of studies on the ecosystem services provided by mountain lions on one hand and still have fingers to spare.

Ironically, the idea that state agencies should continue to be funded primarily by hunters and anglers places state agencies in a catch-22. They are almost always underfunded and desperate for money to address the myriad conservation-management issues of our time, but current politics dictate that they cannot diversify their funding sources, especially from any money with connections to the nonconsumptive public. This has made state wildlife agencies more dependent upon state general funds, and thus more susceptible to the whims of state politicians.[25] There are several current proposals to infuse state agency budgets with

Two female mountain lions squabble over food. Like brown bears, mountain lions growl and hiss when another individual invades their immediate personal space, the size of which appears to be in part defined by the availability of food or size of the carcass. This photograph was part of research funded by private individuals and organizations rather than state funds.

funds to broaden their conservation efforts, including the Restoring America's Wildlife Act, but these proposals serve only to buttress systems in place and not change them.

Instead of seeking alternative sources of money that might threaten current power dynamics, however many state agencies are doubling down in an attempt to increase the dwindling number of hunters instead. It's one of the reasons for concerted efforts to promote hunting and the idea that hunting is at the core of North American conservation efforts. It's the heart of a new movement called the North American Model of Wildlife Conservation (NAM), one that is gaining momentum quickly and being promoted widely by hunters, hunting advocacy groups, and biologists and managers who support a hunting-based model of conservation. In many ways, NAM represents a concerted effort to maintain the status quo and the current position of power held by gun manufactures and sportsman's associations over wildlife management. Should Americans support it, the chances of seeing greater inclusivity in wildlife management, especially management of mountain lions and other predators, will be greatly diminished.

"Since wildlife was financed on a 'user pay' basis, the restoration [of wildlife] fell on the fraction of North Americans who hunt. The rest of society got a free ride in their enjoyment of wildlife as an important component of the high quality of life we enjoy."[26] So wrote biologist, professor, and father of the North American Model of Wildlife Conservation (NAM), Valerius Geist, in a sensationalist 2004 article in *The Outdoorsman*, a periodical read by the hunting community.

Geist was born on February 2, 1938, in Nikolajew, Ukraine, USSR, and raised in Germany and Austria. Since 1977, he has taught at the University of Calgary, where he was the first program director of the environmental science department. Geist is an ardent believer that all landscapes should be managed, and that only under management and through strategic design can sustainable wildlife use and conservation be well achieved. In 1995, he wrote a book chapter in which he articulated his thoughts on a "North American" model of conservation built by hunters and maintained by a hunting legacy. It took several years, as science writing often does, but the idea found support among several key individuals. Geist teamed up with Canadian hunter and biologist Shane Mahoney and US Fish and Wildlife Service biologist John Organ to articulate the NAM more clearly as part of the Sixty-Sixth North American Wildlife and Natural Resources Conference in 2001, an annual conference of the Wildlife Management Institute, a politically influential nonprofit organization with interests supporting the "wise use" of natural resources.

The three-person team was a strategic and effective partnership. Geist proved to be temperamental and was strongly anti-predator. Mahoney, a Canadian hunter and researcher long supported by Safari Club International, was a calm, articulate man who framed his support for the NAM as a model for the conservation of all wildlife, not just game species. Organ was an agency biologist, a professor, a professional member of the Boone and Crockett Club, a life member of the International Hunter Education Association, and soon to be president of The Wildlife Society, among the most influential organizations governing the science driving wildlife management today. Together, the three men brought NAM to the forefront of discussions on state wildlife funding, revive and enlarge the hunting community in the United States and Canada, and conservation more broadly. Together, they built the multifaceted platform, funded by the same special interest groups that lean on state agencies, from which to sell NAM to the people of North America and beyond.[27]

NAM serves two main purposes. First, it rewrites American history, to exaggerate the roles of hunters and hunting in saving North American wildlife, while simultaneously downplaying the contributions of nonhunters. Take for example, the important role of women and nonhunters in ending mountain lion bounties (highlighted in chapter 1). More to the point, it was not hunting but the reduction and regulation of hunting that saved North American wildlife populations. Reading the history as written by the proponents of NAM, one might think that wildlife needs to be hunted to be saved, but that argument is illogical at best.

Second, NAM serves to consolidate power by pronouncing that the tenets of the model already exist and are already embraced by wildlife professionals and conservation scientists more broadly. It's a political maneuver to set forth an alternative narrative, and one that could very well be embraced if it succeeds. Let's consider mountain lion management in the USA to determine whether the tenets of NAM are indeed in operation today.

In brief, the seven proclamations or "pillars" of the NAM, as described by Organ and his team in a Wildlife Society report on the subject,[28] are as follows:

1. Wildlife resources are a public trust.
2. Markets for game are eliminated.
3. Allocation of wildlife is by law.
4. Wildlife can be killed only for a legitimate purpose.
5. Wildlife is considered an international resource.
6. Science is the proper tool to discharge wildlife policy.
7. Democracy of hunting is standard.

The first and third tenets of NAM reflect the legal standing of wildlife ownership and stewardship in the United States and Canada. No one owns wildlife and governments maintain and care for wildlife held in public trust. The third tenet, however, assumes democratic input into wildlife policies and equitable access. The anger and frustration many feel in the cougar conundrum is evidence that equitable access to decision-making is lacking. Critics of the NAM have made clear that the NAM narrative de-emphasizes the historical contributions of women, minorities, and urban communities in North American conservation, and that embracing it will serve to marginalize these communities even further.[29]

The increase in ballot initiatives aimed to stymie state wildlife agency agendas is further evidence for constituents being disenfranchised by their inability to participate in state wildlife management. Twenty-four states provide opportunities for wildlife ballot initiatives as a form of "direct democracy." On the one hand, ballot initiatives provide a voice for those excluded and dissatisfied with state wildlife agencies, but on the other hand they can be used as a strategic means of manipulating public opinion on a wildlife subject. The Humane Society of the United States, for example, was widely accused of using a ballot initiative built on an emotional argument to end mountain lion hunting in Arizona in 2017. Nevertheless, if they had succeeded in getting the legislation on the ballot, it may have passed because most people do not support hunting for reasons other than securing food.

The second tenet of NAM proclaims that wildlife cannot be killed and sold for profit. Fur trapping is in clear violation to the second tenet, as most fur trapping feeds international markets maintained by foreign interests. Proponents of the NAM somehow step around this fact by claiming that fur management is well managed. At this time, the international fur trade does not impact mountain lions, but the growing trade in large-felid body parts to feed Asian markets could impact mountain lions in the near future.

Current mountain lion management definitely violates tenet four: "Wildlife can only be killed for a legitimate purpose." Part of the problem is that the definition of "legitimate" hasn't been agreed upon by all stakeholders, even among hunters themselves. Take for example, "fair chase," which is the set of principles developed by sportsmen hunters more than a century ago to enforce a conservation ethic among its practitioners. Fair chase was originally created by Teddy Roosevelt and the wealthy elite of his era to promote "civility" in killing animals, and to make hunting more palatable for the American public. Perhaps more importantly, fair chase was a strategic means of separating "good" sportsmen hunters from those whom they considered a grave threat to wildlife because they hunted for meat or to bring animal parts to market. Sportsmen of the late nineteenth century promoted the idea that ethical hunters only hunted for leisure and pleasure, and for these reasons they enshrined the idea that fair chase requires giving animals a "sporting" chance to escape.

If one considers Kellert's three types of hunters discussed in the last chapter, fair chase should only appeal to one class of hunters—the dominionistic, and then only if it's forced upon them. Logically, utilitarian hunters who hunt primarily for food should overlook fair chase in favor of any means of acquiring meat more easily. Fair chase might be even more foreign for naturalistic hunters, who hunt primarily to connect with nature. The idea that killing animals is a leisure activity that one enjoys runs counter to everything that naturalistic hunters claim to value.

Today, the definition of fair chase varies by hunter and sportsmen association, but most agree with that preached by the Boone and Crockett Club: "Fair chase is the ethical, sportsmanlike, and lawful pursuit of free-ranging wild game animals in a manner which does not give the hunter an improper or unfair advantage over the animal."[30] The problem, of course, lies with the interpretation of "unfair advantage." Most hunters agree that the animal cannot be restrained, trapped, or found helpless, and importantly that the animal has a "reasonable chance of escape."[31] But some organizations, like Safari Club International, are more liberal in their interpretations.

With respect to mountain lions, the real question is whether hound-hunting violates fair chase guidelines. Even with the best-trained hounds working in ideal conditions, there is never a 100 percent guarantee that a hunter will see and be able to kill a mountain lion, and so it meets the requirements as defined above. Certainly, circumstances can favor the hunter, reducing the chances that the lion will escape, but most of a hunter's advantages are directly proportional to the time the hunter invests before the hunt. First, a hunter needs to be able to identify mountain lion tracks and to approximate the age of footprints they find

An intimate portrait of an adult male mountain lion, the type prized by hunters who seek larger male trophies.

in the field. If hunters haven't learned to identify mountain lion tracks, they'll spend their days trying to sort out the more numerous footprints of deer, elk, coyotes, bobcats, and more. And if they haven't learned to differentiate fresh from old footprints, their hounds may never even catch up to a cat. Second, a dog's ability to follow the scent of a mountain lion relies upon practice and experience—houndsmen must train their hounds and keep them practicing so that they are ready when they encounter a track fresh enough to follow.

The ethics of hound-hunting are increasingly murky, however, because of new technology. Nowadays, houndsmen outfit their dogs with GPS collars that send out blips and bleeps so that they can track their hounds when they move beyond their field of view or hearing. Hunters can follow their dogs in real time on a little screen from the warmth of their truck. This new technology even includes a trigger that alerts the hunter when a hound raises its head to begin barking at the base of a tree—thus hunters can determine when and where their hounds have treed a mountain lion without ever leaving the road where they started.

Sportsmen's associations have had to evolve quickly to address new dog-collar technology, auditory and olfactory calls that attract animals, food attractants like bait piles or salt licks, drones, cellular trail cameras that send images of animals instantly to hunters' phones, and two-way radios used to coordinate hunting efforts, constantly refining their definitions of what "fair chase" really is and what violates ethical hunting. It's a slippery slope for sure, especially given the decline in hunting interest and the constant criticisms from an ever-increasing majority of the American public.

Safari Club International offers zero guidelines on the use of technology in mountain lion hunting, which is not surprising given that they also support hunting in pens (as long as there is suitable cover in which the animal can hide). The Boone and Crockett supports the use of technology "in a way that does not diminish the importance of developing skills as a hunter or reduces hunting to just shooting." They also support the use of VHF and GPS collars to locate hounds after a hunt, but not to aid the hunter in taking a shortcut to the tree. The Pope & Young Club uses the strongest language, stating that hunters can't hunt mountain lions "by the use of electronic devices for attracting, locating, or pursuing game or guiding the hunter to such game."

In short, the Boone and Crockett Club and Pope & Young Club do not believe it is fair chase for hunters to use telemetry or GPS collars to aid them in reaching the tree to shoot a mountain lion, but that the use of new technology is fine as an aid in retrieving dogs if they become lost. But it must be tempting to be halfway up the mountain, sweating and cursing while navigating crisscrossed logs, to have something in one's pocket that could help a hunter save time and navigate the terrain more efficiently.

The second flaw with NAM's fourth tenet and the idea that current mountain lion management is "legitimate" is that only half of all the states that permit mountain lion hunting require that hunters take the meat. Whereas American support for hunting for food remains strong, their intolerance for hunting animals for heads and other trophies continues to grow. Many would argue that killing a mountain lion for its skin is not a "legitimate" reason to kill it. In fact, Organ and his team in their Wildlife Society report about NAM explicitly state that legitimate sportsmen "will not waste any game that is killed."

Tenet 6, that science is the proper tool for discharging wildlife policy, is clearly compromised. Even authors of the Wildlife Society report—NAM proponents—admit that "funding has been largely inadequate to meet the research needs of management agencies, and a trend toward greater political influence in decision-making threatens this principle."[32] As we have discussed over and over in this book, mountain lion management continues to be influenced by special interest groups.

Tenet 7 is perhaps the most disturbing, because it legitimizes the political alliances as well as the money and power issues that are so problematic of wildlife management today. The Wildlife Society report states: "Without secure gun rights, the average person's ability to hunt would likely be compromised, along with *indispensable sources of funding* [emphasis added] for implementation of the Model." Let's be completely clear—all recent fights over gun rights have *not* threatened American rights to own firearms for hunting. This link between hunting rights and broader gun rights was a political coup orchestrated by the NRA following the launch of its lobbying branch, the Center for Legislative Action, in 1975. The NRA's expanded role in politics made the organization more dependent upon corporate sponsors to meet its growing financial needs.[33] Thus the NRA became inextricably tied to the gun manufactures that were filling their coffers, and became completely inflexible on gun-rights issues. Simultaneously, the NRA subverted hunting culture to meet its own political ends. According to academic historian Daniel Herman, hunter conservationists of the Teddy Roosevelt era primarily identified with natural history knowledge as their core value of sportsmanship, but that the NRA has successfully changed the core value of today's hunters to gun ownership instead.[34]

Many now argue that the NRA has become an obstacle rather than an aid to hunters and hunting. One hunter journalist wrote that the NRA is "funding the war on our public lands, while making our beloved sport [hunting] look like a bastion of far-right crackpots."[35] The NRA's divisive language fuels Internet echo chambers and contributes to the greater polarization of wildlife management rather than serving to bridge communities of stakeholders with different perspectives. "To save hunting, you must understand the terms of the battle," writes the NRA. "Because the animal-rights extremists fighting to destroy hunting have an even more destructive goal: the systematic diminishment of humanity itself."[36]

The North American Model of Wildlife Conservation is our past, but it should not be our future. The hunter conservation era is a rich and inspirational heritage and one of which all hunter-conservationists should be proud. But to embrace NAM is to go backwards to a time when wealthy, politically connected white men who abhorred large predators largely decided the fate of North American wildlife, and to ignore the countless essential conservation contributions made by so many others. Any modern conservation model needs to be more inclusive and better reflective of the diversity of ideas, cultures, and values of today's American culture. Without doubt, hunters need to be part of future conservation management. But the management and conservation of mountain lions and our country's broader ecological health and resilience require input from every invested stakeholder.

There are many wonderful, progressive-minded people working for state agencies, and we must be careful not to confuse people with the institutions for which they work, or with the political governing bodies that sometimes dictate their priorities and approve their funding. State agencies are also fantastic at protecting habitat, to the benefit of all species. There is, however, institutional inertia that is resistant to change and resistant to increasing inclusivity and the diversity among state wildlife constituents.[37] This problem, however, is not one-sided. Nonhunters must carry some of the blame for being excluded from discussions and decisions about mountain lion management. Nonhunters are, at times, equally divisive in their communication about people with different values and perspectives. Gary Yourofsky, founder of Animals Deserve Adequate Protection Today and Tomorrow, is perhaps best known for his "eye for an eye" perspective on people who abuse animals. In 2008, he stated, "Deep down, I truly hope that oppression, torture, and murder return to each uncaring human tenfold! . . . Every woman ensconced in fur should endure a rape so vicious that it scars them forever."[38] He went on to speak about the horrors he wished to visit upon men who wear fur, matadors who kill bulls, circus performers who work with captive animals, and scientists who catch and study wildlife.

The simple truth is that most nonhunters are excluded from mountain lion management not because state agencies bar them from participation, but because they do not engage with the process. As seen at the start of chapter 5, Justin Clapp presented proposed changes to mountain lion hunting regulations to the Wyoming Game Commission. Part of his presentation was devoted to reviewing public comments received on an earlier draft of mountain lion hunting regulations, which were weighed while his team finalized proposed mountain lion quotas for the state. One hundred thirty-five individuals and organizations invested the time to engage and comment. In an era when social media sends information alerts around the globe in an instant, this number seems surprisingly low. The Cougar Fund, a national mountain lion advocacy organization, and Panthera, a science-based wild cat conservation organization, both worked in Wyoming at the time, and both organizations encouraged their combined social-media audience of 20,000-plus to send comments to the Wyoming Game and Fish Department.

Ninety-five of the 135 comments were made by hunters, houndsmen associations, and outfitters, and 40 comments were made by nonhunters. It's likely that between the Wyoming staff of the Cougar Fund and Panthera, and their close friends whom they coerced into commenting, that this latter number could be halved—perhaps 20 comments were made by people who do not hunt mountain lions and are unassociated with core mountain lion conservation staff. On the one hand, this tiny number speaks to the apathy of our nation as a whole, and on the other, it highlights why there is biased community engagement with state wildlife agencies. Most nonhunters who value wildlife do not choose to engage or do not know how to engage with their state and federal wildlife agencies. For example, most people in Arizona do not even know the name of the agency that manages their wildlife.[39]

By contrast, houndsmen generally know their local wardens and state wildlife biologists by name. They remain involved in discussions about mountain lion management because they attend public meetings about mountain lion management and submit public comment on mountain lion hunting regulations. Nonhunters who want to voice their opinions about mountain lion management need to jump in and join the fray.

They need to attend public meetings on mountain lion management. They need to get to know their regional state wildlife staff. They need to phone and call their state representatives and share their opinions. They need to support mountain lion research and conservation through donation or volunteering their time.

"*Good* wildlife management is not simply exercising authority over, steadfastly retaining control of, or even taking sole responsibility for wildlife resources; good management is wisely managing the sharing of responsibility for wildlife conservation with stakeholders."[40] We need deer hunters and wildlife advocates to come together to work on mountain lion conservation. We need houndsmen and wildlife biologists. We need wildlife managers and conservation biologists. We need agency staff and representatives from all walks of life to speak with one another and respect each other's values and perspectives. We not only need additional money for state agencies, we need *every* stakeholder to invest money in state wildlife agencies so that agencies recognize every stakeholder equally. We also need state wildlife agencies to use the suite of tools available to them, whether structured decision making as employed in Montana, or conflict resolution, in cases where groups get stuck, to facilitate greater public participation in decision-making with regard to safeguarding mountain lions and other natural resources held in public trust. And if some people or organizations prove resistant to supporting greater inclusivity, we must speak out, rock the boat, and remove them from positions of power.

CHAPTER 9

Toward Coexistence with Mountain Lions

On the 26th of June, 2019, a family of five hiked the trail running between the Laguna Amarga entrance to Torres del Paine National Park and Lake Sarmiento in southernmost Chile. They were led by a guide from the Explora Hotel, and planned to visit the ancient hand paintings made under rock overhangs in the area. While en route to the paintings they encountered fresh mountain lion tracks in the snow, which they stopped to discuss and appreciate. At the time, the mother and her seven-year-old son were hiking slightly ahead of the remainder of the group, and the mother turned back to see the footprints. Simultaneously, movement on the other side of the nearby fence betrayed the presence of a young mountain lion, perhaps the very one that had made the footprints that the group admired.

The mountain lion, which the guide estimated to be six–seven months old, leapt the fence in a graceful bound and dropped into the trail immediately in front of the boy. The boy remained in place, and the guide and his parents spoke to him and told him to stay calm and not to run. The guide moved toward the boy, quickly closing the gap between them, but the boy began to turn back toward the group. The guide told the boy to remain still, but he kept turning, too frightened to listen.

What happened next took everyone's breath away. The mountain lion rose up on its hind feet and placed its paws on the boy's shoulders, in what might have been interpreted as a hug. The boy stood strong, but uttered a single squeak and then began to cry. Perhaps surprised by the noise, the mountain lion dropped to the trail and retreated a short distance away. In that moment, the guide grabbed the boy and confronted the lion. The animal retreated farther from the group. The boy's mother was there the next instant, and after a thorough examination, the group determined that the boy had suffered no injury and that the cat had not extended its claws. The boy, however, was terrified.

The group took some time to re-collect itself and calm the children. Once tensions had dissipated, they began to make their way down the trail. When they looked back, another mountain lion was visible some distance behind them. They thought it might be the mother of the kitten that had embraced the boy. She looked at the group and then moved off in the direction of a group of guanacos, the wild cousins of llamas which are the principal prey of mountain lions in that part of the world.

The Corporación Nacional Forestal (CONAF), Chile's organizational equivalent to the US National Park Service, closed the trail immediately. Before reopening the area to tourists, they erected signs at both ends of the trail dictating rules for those who wanted to hike there: Hikers must remain on the trail and be accompanied by a Park-certified guide. Children younger than fifteen years old must be accompanied by a parent. All groups must register with the Park warden at the southern end of the trail before they set out to hike. The Park did not pursue the young mountain lion that touched the boy, and, we assume, it still wanders the area. In the United States, it would have almost certainly been killed.

Every local guide in Torres del Paine who brings tourists out to see mountain lions has at least one story of an uncomfortably close

An adult female mountain lion and her approximately seven-month-old offspring investigate a camera recently abandoned by a nearby tourist in Patagonia.

encounter. The park administrators live in fear of a mountain lion attack on a tourist. Beyond the tragic implications for the person attacked, such an incident would lead to killing local mountain lions, would destroy local tourism, and perhaps would impact the national economy, given the importance of the park in attracting foreign visitors. Worse, an attack might serve to invite adrenaline junkies who like to place themselves in dangerous situations.

Mountain lion tourism in Patagonia is incomprehensible to most people in the United States—the idea that people walk unprotected with wild mountain lions. But it is important to ponder because it represents an alternative way of living with large carnivores—and very different from our current approach in the United States. Yet even in southern Chile, there are conundrums distinctive to the region. For example, people are beginning to ask: How close is too close to a wild mountain lion? Guides are beginning to wonder whether tourism is impacting mountain lions by reducing the lions' hunting opportunities, which might lead to the starvation of kittens or other negative outcomes.

On the 27th of October, 2019, the mountain lion guides of Estancia Laguna Amarga, which is the current heart of mountain lion tourism in the region, called a meeting among themselves to address these very issues. Pushy photographers and filmmakers had influenced guiding in recent years—people were forever trying to get closer to wild mountain lions. Photographers and filmmakers now rush ahead of mountain lions to place themselves in the animal's path, in order to photograph its face while it is walking. Just several years earlier, the guides never saw anyone blocking mountain lions or forcing them to change direction. And so the guides of Laguna Amarga now decided that enough is enough. They agreed to create safety standards to protect mountain lions from people, and to ensure mountain lion tourism is sustainable for years to come.

North American cougar conundrums are different. The people here are a conglomeration of different cultures, and we have a different history of living with mountain lions. Few people in North America, for example, have encountered mountain lions in the wild. Whereas poaching in Chile and provisional bounties paid for mountain lion skins in Argentina threaten mountain lions in Patagonia, it is habitat destruction and habitat fragmentation that are the primary threats to mountain lions in the United States and Canada. Every year, wildlands become houses and shopping malls and natural-gas pads that will never again harbor mountain lions or other wildlife. Highways cut up the largest remnants of wildland, and as traffic increases, they become impenetrable barriers that mountain lions cannot cross. Mountain lions on the Olympic Peninsula in western Washington state exhibit lower genetic diversity than their mainland counterparts,[1] and the fast-growing I-5 corridor connecting Seattle to Portland, Oregon, is turning the peninsula into an island. In 2018, the Skokomish Tribe captured a young male mountain lion on the peninsula northwest of Olympia. He tried three times to cross Interstate-5 before he traveled south, where he failed to cross the massive Columbia River. Then he turned west where his dispersal was stymied by the Pacific Ocean. He was ultimately killed by a hunter before he could establish a home range.

Mountain lions along the coast of Southern California are also suffering genetic deterioration, isolated as they are by a spiderweb of highways and urban sprawl. Since 2002, Seth Riley and Jeff Sikich have been studying the tiny mountain lion population that inhabits the Santa Monica mountain range just north of Los Angeles, as highlighted in the story about P-45 found in the preface. In recent analyses led by John Benson, Riley and Sikich predicted that remnant mountain lion populations in both the Santa Monica Mountains and the Santa Ana Mountains south of LA could go extinct within twelve to fifty years, should they remain disconnected from other mountain lion populations.[2]

The people of Los Angeles, however, are not idly standing by. They are rallying to raise $87 million to build a wildlife bridge over Highway 101 that will allow mountain lions and other animals to move between the Santa Monica Mountains and wildlands to the north. P-22, their local celebrity mountain lion, is the poster–mountain lion for the effort. Fast on the heels of the publication of the aforementioned research, the Mountain Lion Foundation and the Center for Biological Diversity also submitted a formal request for southern, coastal mountain lion populations of be listed as endangered by the California Department of Fish and Wildlife due to their genetic vulnerability. The proposal is currently under review by state biologists.

The logo for Save LA Cougars (savelacougars.org), the primary collaborative effort to raise funds for the wildlife bridge to aid Los Angeles's mountain lions in the Santa Monica Mountain Range. (Artwork copyright National Wildlife Federation. Used with permission of B. Pratt / NWF.)

On a smaller scale, new extraction routes built to remove natural resources continue to cut up public lands, providing hunters and four-wheel-drive vehicles deeper access into the last wild refuges in North America. The ever-increasing number of roads in a remote region are one of the leading influences on the survival of many species there—deer, elk, wolves, bears, and mountain lions among them. Other potential threats to mountain lions include diseases contracted from domestic cats and dogs, injuries from getting caught in traps set for bobcats, plague, poisoning by marine fog carrying deadly mercury, and rodenticides that are so liberally used in many Western states that mountain lions ingest through predation or scavenging poisoned prey. Rodenticide poisoning in particular is of real concern along the West Coast, where it impacts numerous species in interconnected food webs. The extent of these threats on the mountain lion superorganism is not yet well understood, nor, indeed, whether they really *are* threats as defined by the International Union for Conservation of Nature (see chapter 5).

As practiced today in the western United States and Canada, hunting mountain lions is not a threat to the long-term viability of the mountain lion superorganism. Nevertheless, there is evidence that we are over-hunting them. Given the social implications of hunting in a society of mostly nonhunters, and the effects of hunting on the integrity and health of mountain lion populations, hunting will remain a controversial component of the cougar conundrum. To be clear, there are rarely biological reasons to hunt mountain lions, and the removal of mountain lions that attack livestock or people is best done by professionals rather than recreational hunters.

There are, in fact, several reasons why we should limit mountain lion hunting. Most importantly, hunting may increase negative interactions between mountain lions and people and livestock, rather than decrease them. Mountain lion populations do not need to be controlled by human hunting, either, as they are already regulated by prey availability and internal social behaviors. Nevertheless, further research on whether hound-hunting instills fear of people or places, thereby indirectly protecting livestock and people, is warranted, given that it appears to be true for other species like bears. But at this time, there is no evidence that hound-hunting protects livestock or people by making mountain lions more afraid. Even if hound-hunting does instill fear, it may only be fear of dogs rather than people. Many believe that mountain lions are already hardwired to fear dogs because of their coevolution with wolves, which chase and sometimes kill them.

The best reason to maintain any hunting at all—and this speaks specifically to hound-hunting—is that many hound-hunters advocate for the species, and because houndsmen are resources sometimes needed by state agencies and wildlife researchers. This is not to imply that mountain lion conservation requires hunting. California models an alternative approach in which mountain lions are fully protected and yet people still support mountain lions. As a whole, evidence suggests that the people of California suffer no negative consequences for protecting the species when comparing the various metrics of deer density and human safety to other states in which hunting in permitted.[3] That is despite consequences due to the scarcity of hound-hunters as a resource, voiced by the state wildlife agency and mountain lion researchers who need them.

Conservation can also occur simultaneously with hunting. Where hunting is established, for example, we can improve management so as to maintain the benefits of hound-hunters, while simultaneously ensuring healthier mountain lion populations and less conflicts with people. First, we need to kill fewer mountain lions so as not to disturb their age and sex structures so dramatically, and to ensure that mountain lion populations function naturally in terms of their social interactions,

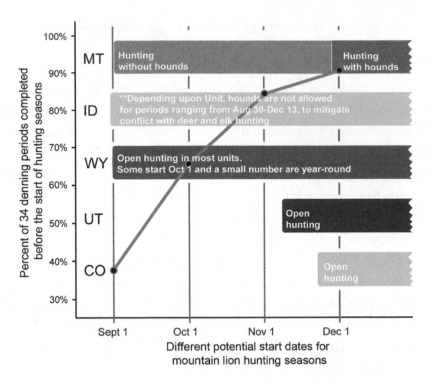

This graphic depicts the 2017 mountain lion hunting seasons for five Western states, and the percentage of thirty-four dens monitored in northwest Wyoming that were completed at the start of each month, running from September to December. For example, delaying the hunting season until December 1 would avoid >90 percent of denning females, and give the best opportunities for hunters to protect females with dependent young. (Originally published in O'Malley et al., "Aligning Mountain Lion Hunting Seasons to Mitigate Orphaning-Dependent Kittens," fig. 4.)

dispersal patterns, and ecological contributions to ecosystems upon which we depend. The easiest way to achieve this is to reduce hunting quotas. Rich Beausoleil, bear and cougar specialist for the Washington (State) Department of Fish & Wildlife, and his team have come up with another solution as well—redistributing mountain lion hunting by managing more numerous, smaller game units with smaller hunting limits. The effect of this approach mimics reducing quotas in that it stops hunters from removing too many mountain lions from a single population, and subsequently causing adverse juvenile-delinquent effects.[4]

Second, we need to better protect females and kittens from unintentional impacts of hunting. For example, in northern climates with clear seasonality, mountain lions exhibit a birth pulse in summer and early fall, and thus delaying the hunting season until the first day of December substantially decreases the number of females wandering around while their young kittens are hidden away in dens.[5] Delaying the hunt provides hunters the best opportunity to identify females with young via footprints made in snow, so that they do not accidentally kill them. Alternatively, more Western states could impose lower limits on the number of females they permit hunters to kill each year.

Third, we need to invest in new research to understand the complex indirect effects of hunting on mountain lion social behaviors and population dynamics, as well as other effects we have yet to understand. This will be time-consuming, expensive research. With it, however, we will better understand the ecological consequences of hunting-based management. This is as crucial to the defense of hunting mountain lions as it is to gathering evidence to challenge it.

We can also improve hunting practices. All state wildlife programs should educate hunters in differentiating male from female mountain lions, as those states that provide training show that hunters can correctly differentiate between the sexes more often than not.[6] States should also require hunters to carry out and utilize the meat of the mountain lions they kill. Americans are decidedly against trophy hunting, but generally do support hunting for food.[7] Hunting for skins and skulls also encourages dominionistic hunting that typically undermines strategies to peacefully coexist with mountain lions, as well as the conservation of wildlife more broadly.

"The Klickitat County Sheriff's Office and other Law Enforcement Agencies in Klickitat County has received numerous complaints in regard to Cougar sightings in Goldendale, White Salmon, Glenwood, Husum, and other neighborhoods," wrote Sheriff Bob Songer for posting on the county Facebook page. "This is a serious PUBLIC SAFETY CONCERN. In addition to the Public Safety Concern, cougars and bears are a major concern and a problem for the farmers and ranchers in our County. As Sheriff, I have decided immediately to establish a program in accordance with Washington State Law, (RCW 77.15.245(2a), which will allow better protection for the Public Safety, the protection of domestic animals and livestock within and throughout Klickitat County."[8]

Alerted by a rancher that such a legal loophole existed, the sheriff took charge of mountain lion management in Klickitat County, Washington, in August, 2019. Whereas state wildlife agencies generally balk at ballot initiatives that are run by wildlife advocates and that undermine them, this was a threat to their authority brought forth by people they thought of as their constituents. Sheriff Songer deputized a "posse" of five local houndsmen in a state in which hound-hunting is currently illegal. They respond to sightings of mountain lions reported by the public, with the authority to chase and kill them. Since Songer's announcement, the Klickitat County Sheriff's Office Facebook page has become host to bullying and misinformation. Hunters, empowered and legitimized by an authority, jeer and insult wildlife advocates and others opposed to the actions of the sheriff and his posse.

"We don't have to wait for a killing," Songer explained, referring to a livestock depredation. "I feel very strongly that prevention is better than waiting for something to happen."[9] Three months into the change, Songer's posse had killed eight or nine mountain lions,[10] which is 66 or

75 percent of the annual hunting limit regulated and managed by the Washington Department of Fish and Wildlife for game units 578, 388, and 382—that is, Klickitat County. This year, however, the state limit of twelve mountain lions could be killed in addition to the unlimited number of cats killed by the sheriff's posse.

When the Washington Department of Fish and Wildlife was asked for comments about local law enforcement taking management into their own hands and hunting mountain lions, they responded that they had "no opinion." Caught between a rock and a hard place, how could they openly oppose another government agency? The actions of the Klickitat County sheriff's office represent an unprecedented attack on state wildlife management—now there are three county sheriff's offices managing mountain lions in Washington—and one for which most statewide agencies remain completely unprepared. Nevertheless, it underscores the importance of democratizing decision-making when managing mountain lions, or else people who feel disenfranchised with current mountain lion management will continue to work to undermine state wildlife agencies in any way they can.

Participatory, collaborative decision-making is especially essential to quelling growing concerns over the disproportionate representation of hunters and people with anti-predator sentiments in state wildlife commissions and more broadly in decision-making. Social scientists suggest that for any group to feel "legitimate," there must be "fair representation, appropriate government resources, and a consensus-driven decision-making process."[11] Thus, change should begin at the top, where politics influence how wildlife priorities are established, as well as in the selection of state wildlife commissions, which approve and set state wildlife regulations. Between 1987 and 1997, for example, the number of states in which governors or state cabinet members, rather than wildlife personnel, establish state wildlife priorities increased from six to twenty-seven.[12]

Governors also need to ensure that their appointed commissions include the full spectrum of interests from all of their constituents— or alternatively, commissions need to become elected positions so that they do so.[13] Specifically, wildlife commissions need representation from among wildlife watchers, the ecotourism industry, and other types of nonconsumptive users. Once these opportunities are afforded them, nonhunters need to be ready to seize them. This means nonhunters need to engage with mountain lion management much more.

State management agencies also need to host opportunities for public participation, and be willing to share decision-making power in order to ensure that people have the incentive to attend. Ballot initiatives meant to undermine state wildlife management agencies are democratic in that they allow for broader participation, but they fall short on bringing people together to collaborate and negotiate solutions. Ultimately, the democratic component of ballot initiatives should instead be preemptively included in state wildlife management decisions through various participatory processes and the conflict mitigation tools available to them, as role-modeled by the Montana Department of Fish, Wildlife & Parks, Mike Mitchell, and lion and elk hunters in Montana's Region 2. Ideally, state wildlife managers will be proactive in initiating greater public input and the evolution of their organizations, and thus become positive agents of change.[14]

We need bridges, not divisions, among stakeholders. Houndsmen and nonhunters need to stop flinging mud at each other on social media, and recognize each other as allies instead. They need to listen to each other and realize that all people have equal rights, even if their values and perspectives are different from their own. We need to remember that many of the perspectives we encounter on social media represent extremes, and not the opinions of the majority of people, which lie somewhere in the middle. We must stop seeing people as us-versus-them, and recognize that aiding mountain lions is far more important than winning a skirmish with the opposition. Not all mountain lion lovers are anti-hunting, anti-guns, liberal, vegetarian city slickers who drive Priuses, and not all hunters are conservative ranchers who kill African lions in canned hunts, drive big trucks, drink cheap beer, and shoot anything that moves just for the fun of it. The choice is ours as to whether we are faced with a conundrum or an opportunity. We can despise those who are different from us, or learn to communicate and work with them toward shared goals.

Perhaps most importantly, we need to change the revenue streams supporting state wildlife agencies. Yes, we need to support new funding for state wildlife programs, as currently proposed by the Restoring America's Wildlife Act, but this is not enough to shift agency paradigms from one that prioritizes paying customers to one of protecting the interests of all people equally. We need every interested stakeholder, from butterfly watcher to elk hunter, to pay into state wildlife programs in order to change how state agencies operate. Maybe it is time to propose Teaming with Wildlife legislation a second time, this time led by the Conservation Alliance, a 235-member organization of outdoor gear companies founded by Patagonia, REI, Kelty, and The North Face, and dedicated to conservation. Or perhaps the Outdoor Industry Association's Sustainability Working Group, dedicated to seeing outdoor gear made and supported by ecologically and socially sustainable means. Americans have changed and many are willing to aid wildlife voluntarily by shouldering a small tax on outdoor gear. The question is whether the outdoor gear corporations have changed enough to join them in supporting such legislation.

Last, we need to reintroduce mountain lions to everyday people. We need to get our facts straight and describe the real animal, not the snarling beast surrounded by hounds, or the coldhearted killing machine of myth and legend. We need to recognize that real mountain lions are individual characters, and not all the same. We must harness photography and film, as they have proven powerful tools in reintroducing mountain lions to broader audiences, as well as motivating people to participate in their conservation.

Strategic research that fills gaps in our knowledge about mountain lion behaviors and other natural history is equally compelling and engaging. Even after sixty years of research on the species, we have so much to learn about their mostly secret lives. Nevertheless, conservation and mountain lion advocacy organizations also have their work cut out for them in terms of inspiring people to participate and to remain invested in state mountain lion management. Adding frown emojis to posts about unjust or unethical management practices is not participating, it's spectating. We need people to engage.

M68, the young male mountain lion and one-time sheep killer mentioned briefly in chapter 3, dispersed 100 miles south of Jackson, Wyoming, to set up a territory deep in the Wyoming Range. It was inaccessible terrain in general, soon to be cut off completely by winter snows. Panthera's Teton Cougar Project decided to be proactive and remove M68's collar while it was working, and GPS data made determining his whereabouts so easy. The team called Boone Smith, who had been the houndsman for the Teton Cougar Project since its inception in 2000.

Subsequently, three generations of Smiths joined Panthera in search of M68 on a chilly day in late October. Sam Smith, Boone's father, humble, generous, and sturdy. Boone Smith, athletic, capable, and forever positive. Jake Smith, Boone's son, entrusted to Grandpa should the team move too quickly. The Smiths were essential team members in more than catching mountain lions. They mentored field staff in mountain lion natural history and field skills, collected field data, saved field

Mountain Lions Increase an Ecosystem's Health and Biodiversity

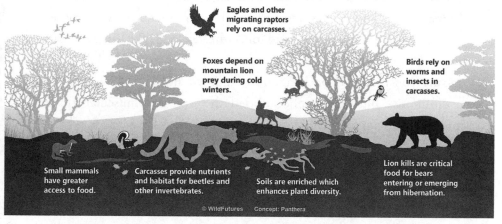

Eagles and other migrating raptors rely on carcasses.

Foxes depend on mountain lion prey during cold winters.

Birds rely on worms and insects in carcasses.

Small mammals have greater access to food.

Carcasses provide nutrients and habitat for beetles and other invertebrates.

Soils are enriched which enhances plant diversity.

Lion kills are critical food for bears entering or emerging from hibernation.

© WildFutures   Concept: Panthera

A graphic co-created by WildFutures, a mountain lion education and advocacy organization, and Panthera, a science-based conservation organization, depicting some of the many positive effects of mountain lion predation within their larger ecological communities.

staff from themselves when their trucks and snowmobiles got stuck, and helped decide how best to invest research time and effort.

Biologists and houndsmen have collaborated to study mountain lions since the very first mountain lion project began in central Idaho, when houndsman Wilbur Wiles teamed up with then student biologist, Maurice Hornocker. Over sixty years of research, both houndsmen and biologists have benefited from the exchange of experience and knowledge shared by the other. So crucial is the houndsman's role in mountain lion research that California biologists, suffering the loss of the skill set following the ban on hound-hunting, solicited support from the California Department of Fish and Wildlife for an exemption that would allow select houndsmen to practice chasing mountain lions. In 2019, the state of Washington passed legislation to do just that—to grant the state wildlife agency the authority to permit select houndsmen to practice and train their hounds, so that they could stand ready to assist the state in the case of livestock depredation or, worse, an attack on a human being, and to assist researchers who might require their services.

After navigating dirt roads and four-wheeler trails, the Panthera team located M68 where he was lying up near a young elk he'd killed and mostly consumed. Boone's hounds exploded into action. Loud Sue, a diminutive bluetick coonhound, led the way, and muscular Trailer and lean Poncho followed close on her heels. Sue's long, bawling barks betrayed their pursuit and their direction of travel, and the team stood and listened as the hounds settled in to work the trail. Then they began to follow, but it was only a short hike before Sue's voice switched to shorter, repetitive, punchy bawls. Soon the team heard Trailer's low, gruff barks, too, mixed with Poncho's barks. The change in their barks indicated that they were sitting somewhere nearby, looking up at M68.

Once hounds "jump" a mountain lion—meaning that the hounds catch up to the cat, and the cat becomes aware of the threat and bolts—the chase is generally short, at least in forested environments. Dr. Chris Wilmers and his research team measured the energetic demands of mountain lions while they were being pursued by hounds, and found that they were negligible. The average chase in the densely forested biome near Santa Cruz, California, where his team worked, was only 3 minutes 45 seconds.[15] Although the mountain lion's energy expenditure while fleeing was five times higher than when hunting, the short duration of the expenditure meant that it was a small energetic cost by comparison to the energy expended over a typical day. Perhaps the energy expended during a chase could be thought of as someone taking a thirty-minute jog in the morning. In the more open or clumped forests of the drier Rocky Mountains, however, hound chases tend to be longer and may have a greater impact on mountain lions. Certainly, savvy lions learn to fool chasing hounds by running in loops, jumping between trees, and numerous other strategies. These mountain lions tend to run longer and expend much more energy in doing so.

The team closed in quickly on the barking hounds. When they could see M68 in the tree, they stopped to prepare a dart that would anesthetize the big cat for safe handling. Everyone knew their roles and the work was smooth. M68 was shot with an CO_2-powered rifle, adjusted down to minimize the impact of the dart. Then he was lowered safely to the ground, where the team was ready to receive him. They checked and monitored his temperature, breathing, and heart rate while they worked, to ensure that he was well. They took several measurements to document his growth. He weighed 142 pounds, as compared to 101 pounds just one year earlier.

M68 was healthy, robust, and beautiful. Even though he barely made any kills the previous winter, he had survived three full months on contributions provided by several adult lions that allowed him to feed from their kills. But as winter hinted at spring, he came into his own and began to support himself. The team couldn't help but marvel at M68's transformation.

Then they removed his collar, and gave him one last injection, this one meant to reverse the effects of anesthesia. The group gathered expectantly to one side. Several minutes later, M68 rose slowly and turned to have one last look at the team. He'd grown into an impressive animal with a radiant coat and the intent stare distinctive of mountain lions. Given another year or two, he'd become larger and more imposing still. Then, as if realizing his situation, he fled the group, disappearing momentarily in the lattice of branches provided by a recently fallen tree. There, he paused to reassess. He reappeared a minute later on the far side of the woodpile, walking rather than running.

"Wow," Boone said smiling, breaking the silence. "That's the first time I've done that. It feels good to take a collar off a cat instead of put one on."

Male mountain lion, northern Rockies.

Sam nodded, smiling just as broadly. Jake beamed by his side.

"Indeed it does," I said in response, feeling lucky to witness a wild mountain lion walking free. Happy, in fact, to participate in the cougar conundrum. M68 disappeared around the corner of the hillside without another backwards glance.

"It is best to be both feared and loved; however, if one cannot be both it is better to be feared than loved," wrote philosopher and politician, Niccolò Machiavelli many centuries ago.[16] That might be good advice for dictators, but it clearly hasn't worked for mountain lions. Our fear of mountain lions has fed the mythology perpetuating their persecution for hundreds of years. Today, however, security cameras monitoring our entranceways and backyards as well as trail cameras set by hunters and other wildlife enthusiasts in nearby woods provide irrefutable evidence that mountain lions live among us like ghosts. The cougar conundrum is in fact an amazing opportunity to witness the restoration of a native carnivore to former range. It's an opportunity to improve our continent's ecological health and resilience, and to build bridges among people from different walks of life so that we work together to protect our shared natural resources. Both humans and mountain lions are part of natural ecosystems.

To foster peaceful coexistence between our two species, we must learn to live differently in the presence of mountain lions. We must pay more attention to our surroundings. We must keep an eye on our pets, our livestock, and our children. Livestock owners, too, need to be held more accountable for the animals they own. This is especially true for people with small herds or hobby animals. People who do not protect their animals from predators should not be issued depredation permits if they suffer losses. They should not be granted support in killing mountain lions, whether via state-funded houndsmen, state wildlife personnel salaries, or the gas that goes into the trucks used to pursue the perpetrating animal, unless they themselves have invested in building the infrastructure to protect their animals appropriately. If they cannot afford such investments, state agencies, conservation organizations, and mountain lion advocacy groups need to help them, as did the National Wildlife Federation and Mountain Lion Foundation in the story of P-45, found in the preface of this book.

Some people will argue that the changes prescribed above are unnecessary burdens, and that it would be easier to live without mountain lions. What they do not recognize is how much mountain lions enhance our lives. Mountain lions are evidence that human beings can protect and bring a species back from near extinction. They are evidence of true wilderness: graceful, majestic, and strong. Mountain lions are keystone species that enrich and strengthen the forests, mountains, and desert ecosystems upon which we depend—supporting mountain lions, in turn, supports our own health and communities.

It is our decision as to whether we will coexist with mountain lions. As individuals, we can make choices that facilitate coexistence in our own backyards. As a society, we can work together to design wildlife conservation management that benefits mountain lions and healthy ecosystems for all people. Regardless of what we decide, we can be sure that mountain lions will continue to pursue their own best interests. Their intelligent, resilient capabilities are in fact good news for us, because it makes living with healthy mountain lion populations very straightforward. We need only choose to share the world with them rather than persecute them, and they will take care of the rest.

Notes

Preface

1. Steven G. *Torres* et al., "*Mountain Lion and Human Activity* in California: Testing Speculations," *Wildlife Society Bulletin* 24, no. 3 (Autumn 1996): 451–60.
2. Kevin Hansen, *Cougar: The American Lion* (Flagstaff, AZ: Northland Publishing, 1992), 58.
3. Anna Vecchio, "Wildlife Biologist Miguel Ordeñana Joins National Wildlife Federation Board of Directors," National Wildlife Federation, June 18, 2019, https://www.nwf.org/Latest-News/Press-Releases/2019/06-18-19-Miguel-Ordenana.
4. Dana Goodyear, "Lions of Los Angeles: Are the City's Pumas Dangerous Predators or Celebrity Guests?," *New Yorker*, February 5, 2017, https://www.newyorker.com/magazine/2017/02/13/lions-of-los-angeles.
5. Adam Popescu, "What It's Like Living in California's Mountain Lion Country," BBC, December 19, 2017, http://www.bbc.com/future/story/20171218-what-its-like-living-in-californias-mountain-lion-country.
6. Associated Press, "California Adopts Three-Strikes Policy for Marauding Mountain Lions," CBS Los Angeles, January 8, 2018, https://losangeles.cbslocal.com/2018/01/04/california-three-strikes-mountain-lions/.
7. Theodore Roosevelt, "With the Cougar Hounds," in *Outdoor Pastimes of an American Hunter* (New York: Scribner's Sons, 1905).

Chapter 1

1. Stephen Demarais, Karl V. Miller, and Harry A. Jacobson, "White-Tailed Deer," in *Ecology and Management of Large Mammals in North America*, eds. Stephen Demarais and Paul R. Krausman (Upper Saddle River, NJ: Prentice-Hall, 2000), 601–28.

2. Stanley Young and Edward A. Goldman, *Puma: America's Mysterious Cat* (New York: American Wildlife Institute, 1946).

3. Ibid, 167.

4. J. Bob Tinsley, *The Puma: Legendary Cat of the Americas* (El Paso, TX: Texas Western Books, 1987).

5. Ibid.

6. Paula Wild, *The Cougar: Beautiful, Wild, and Dangerous* (Madeira Park, BC: Douglas and McIntyre, 2013), 53–60.

7. From the cover of Jay C. Bruce, *Cougar Killer* (New York: Comet Press Books, 1953).

8. Frank Dobie, *The Ben Lilly Legend* (Boston: Little Brown and Co, 1950), 61.

9. Wikipedia, "Ben Lilly," accessed November 15, 2017, https://en.wikipedia .org/wiki/Ben_Lilly.

10. Rachel Carson, *Silent Spring* (Boston: Houghton Mifflin, 1962), 99.

11. David Mattson and Susan Clark, "People, Politics, and Cougar Management," in *Cougar Ecology and Conservation*, eds. Maurice Hornocker and Sharon Negri (Chicago: University of Chicago Press, 2009), 206–20.

12. David *Brown*, "*A Lion for All Seasons*," in *Proceedings of the Second Mountain Lion Workshop*, eds. Jay Roberson and Frederick Lindzey (Utah: Zion National Park, *1984*), 13–22.

13. Wain Evans, "New Mexico," in *Transactions of the First Mountain Lion Workshop*, eds. Glen Christensen and Robert Fischer (Sparks, NV: Nevada Fish and Game Department, 1976), 25–26.

14. Josiah Priest, "Wheaton and the Panther," in *Stories of the Revolution* (Albany, NY: Hoffman and White, 1836).

15. Daniel Bryan, *The Mountain Muse* (Harrisburg, PA: Davidson and Bourne, 1813), 142.

16. Ibid., 45.

17. Jay C. Bruce, *Cougar Killer* (New York: Comet Press Books, 1953).

18. Theodore Roosevelt, *The Works of Theodore Roosevelt: Outdoor Pastimes of an American Hunter* (New York: Charles Scribner's Sons, 1906), 51.

19. George Musters, *At Home with the Patagonians: A Year's Wanderings on Untrodden Ground from the Straits of Magellan to the Rio Negro* (London: George Murray, 1871), 53.

20. Bryan, *The Mountain Muse*, 142.

21. Theodore Roosevelt, *A Book-Lover's Holidays in the Open* (New York: Charles Scribner's Sons, 1916), 22.

22. Claude Barnes, *The Cougar or Mountain Lion* (Salt Lake City, UT: The Ralton Company, 1960), 10.

23. Musters, *At Home with the Patagonians*, 52.

24. Aristotle, *Complete Works of Aristotle: The Revised Oxford Translation*, vol. 1 (Princeton, NJ: Princeton University Press, 1984), 1244.

25. James DeKay, *Natural History of New York: Zoology* (Albany, NY: W. & A. White & J. Visscher, 1842), 49.

26. William Byrd, *History of the Dividing Line*, in the Westover Manuscripts dated 1728 (Petersburg, VA: Edmund & Julian C. Ruffin, 1841), 155.

27. Frank Bostock, "Letter to President Roosevelt," in Roosevelt, *Outdoor Pastimes of an American Hunter*, 406–7.

28. Roosevelt, *Outdoor Pastimes of an American Hunter*, 19.

29. Roosevelt, "A Cougar-Hunt on the Rim of the Grand Canyon," in *A Book-Lover's Holidays in the Open*, 16.

30. Ibid, 17.

31. Ronald Douglas Lawrence, *The Ghost Walker* (New York: Holt, Rinehart and Wilson, 1983), 21.

32. Leslie Patten, *Ghostwalker: Tracking a Mountain Lion's Soul through Science and Story* ([no city listed]: Far Cry Publishing, 2018), 84.

33. Linda L. Sweanor, Kenneth A. Logan, and Maurice G. Hornocker, "Puma Responses to Close Approaches by Researchers," *Wildlife Society Bulletin* 33, no. 3 (2005): 905–13.

34. John L. Gittleman, "Carnivore Life History Patterns: Allometric, Phylogenetic, and Ecological Associations," *American Naturalist* 127, no. 6 (June 1986): 744–71.

35. Travis D. Bartnick et al., "Apparent Adoption of Orphaned Cougars (*Puma concolor*) in Northwestern Wyoming," *Western North American Naturalist* 74, no. 1 (December 2013): 133–37; L. Mark Elbroch and Howard Quigley, "Social Interactions in a Solitary Carnivore," *Current Zoology* 63, no. 4 (August 2017): 357–62.

36. L. Mark Elbroch et al., "Adaptive Social Behaviors in a Solitary Carnivore," *Science Advances* 3, no. 10 (October 2017): e1701218.

Chapter 2

1. Tony Perry, "Rangers Kill Cougar after Girl Is Attacked," *Los Angeles Times*, September 19, 1993, http://articles.latimes.com/1993-09-19/news /mn-36917_1_mountain-lion-sightings.
2. Kevin Johnson, "Cougars' Worst Foe Is Found to Be Cars," *Los Angeles Times*, August 9, 1993, http://articles.latimes.com/1993-08-09/news/mn -22025_1_santa-ana-mountains.
3. Associated Press, "Cougar Is Killed for Jogger's Death in California Park," New York Times, May 2, 1994, http://www.nytimes.com/1994/05/02/us /cougar-is-killed-for-jogger-s-death-in-california-park.html; Peter H. King, "The Lion and the Jogger," *Los Angeles Times*, May 1, 1994, http://articles .latimes.com/1994-05-01/news/mn-52729_1_mountain-lions.
4. Justin Nobel, "On Mountain Lions, and the Forgotten Story of the 1909 Church Women Mauled by a Rabid Cougar," *Huffington Post*, March 12, 2013, www.huffingtonpost.com/justin-nobel/on-mountain-lions-and-the_b _2861600.html.
5. Associated Press, "Woman Kills Puma with Knife after It Attacked Camp- ers," *Desert News*, August 17, 1994, www.deseret.com/1994/8/17/19125642 /woman-kills-puma-with-knife-after-it-attacked-campers.
6. Doug Smith, "Too Close to Home: Mountain Lions' Contact with Humans Increases as Their Numbers Grow," *Los Angeles Times*, September 6, 1994, http://articles.latimes.com/1994-09-06/local/me-35233_1_mountain -lion-s-rebound/2.
7. Ted Williams, "The Lion's Silent Return," in *Shadow Cat: Encountering the American Mountain Lion*, eds. Susan Ewing and Elizabeth Grossman (Seat- tle, WA: Sasquatch Books, 1999), 13–26.
8. Rene Lynch and Kevin Johnson, "Cougar Attack Settlement Costs O.C. $1.5 Million," *Los Angeles Times*, October 22, 1993, www.latimes.com /archives/la-xpm-1993-10-22-mn-48511-story.html.
9. David Mattson, Kenneth Logan, and Linda Sweanor, "Factors Governing Risk of Cougar Attacks on Humans," *Human–Wildlife Interactions* 5, no. 1 (Spring 2011): 135–58.

10. Christopher Ingraham, "Chart: The Animals That Are Most Likely to Kill You This Summer," *Washington Post*, June 16, 2015, http://www.washington post.com/news/wonkblog/wp/2015/06/16/chart-the-animals-that-are -most-likely-to-kill-you-this-summer/; Kaiser Fung and Andrew Gelman, "Debunking the Great 'Selfies Are More Deadly than Shark Attacks' Myth," *Daily Beast*, April 13, 2017, www.thedailybeast.com/articles/2015/10/05 /debunking-the-great-selfies-are-more-deadly-than-shark-attacks-myth .html; Jeff Wise, "Survival Tips You Must Know," *Popular Mechanics*, February 1, 2017, www.popularmechanics.com/adventure/outdoors/tips/a3114 /how-not-to-die-20-survival-tips-you-must-know-16030884/.

11. Seth Hettena, "Study Sheds New Light on Cougars, Humans," *NBC News*, April 6, 2004, www.nbcnews.com/id/4677182/ns/us_news-environment /t/study-sheds-new-light-cougars-humans/#.Vi1Ye6ToggU.

12. Aliah A. Knopff, Kyle H. Knopff, and Colleen C. St. Clair, "Tolerance for Cougars Diminished by High Perception of Risk," *Ecology and Society* 21, no. 4 (2016): 33.

13. Cougar Management Guidelines Working Group, *Cougar Management Guidelines—First Edition* (Washington, DC: WildFutures, 2015), 80.

14. Steven G. *Torres* et al., "*Mountain Lion and Human Activity* in California: Testing Speculations," *Wildlife Society Bulletin* 24, no. 3 (Autumn 1996): 451–60.

15. Liza Gross, "The Man Who Made California Safe for Lions," *Mountain Lion Foundation*, October 15, 2012, www.mountainlion.org/featurearticleguest unlikelyally.asp.

16. John F. Dunlap, *Senator from Napa* (website), accessed December 31, 2019, http://www.senatorfromnapa.com/.

17. Tony Perry, "Slain Mountain Lion Is One That Killed Hiker," *Los Angeles Times*, December 13, 1994, http://articles.latimes.com/1994-12-13/news /mn-8439_1_mountain-lion-foundation; Joe Garofoli, "Mountain Lion Attacks Kill 1, Injure 1—Loss of Habitat May Fuel Run-ins, Experts Suggest," *San Francisco Gate*, January 10, 2004, www.sfgate.com/politics /joegarofoli/article/Mountain-lion-attacks-kill-1-injure-1-Loss-of -2831604.php.

18. *Kat Kerlin*, "Researchers Delve Deep into Lion Country," *UC Davis Dateline*, October 19, 2001, *http://dateline.ucdavis.edu/dl_detail.lasso?id=7288*.

19. Jennifer R. Wolch, Andrea Gullo, and Unna Lassiter, "Changing Attitudes towards California's Cougars," *Society and Animals* 5, no. 2 (1997): 95–116.

20. "Cougars Under Fire on California Ballot," *Washington Post*, March 18, 1996, www.washingtonpost.com/archive/politics/1996/03/18/cougars-under -fire-on-california-ballot/f4fed0cf-02df-4f13-8539-e1bcb30a3154 /?noredirect=on&utm_term=.78aa95f37a9e.

21. Bill Karr, "My Turn: Mountain Lion Attack Proves Hounds Beneficial," *Mountain Democrat*, July 16, 2012, www.mtdemocrat.com/opinion/my -turn-mountain-lion-attack-proves-hounds-beneficial/.

22. Valerius Geist, Shane P. Mahoney, and John F. Organ, "Why Hunting Has Defined the North American Model of Wildlife Conservation," *Transactions of the North American Wildlife and Natural Resources Conference* 66 (2001): 175–85.

23. Theodore Roosevelt, "A Cougar-Hunt on the Rim of the Grand Canyon," in *A Book-Lover's Holidays in the Open* (New York: Charles Scribner's Sons, 1916).

24. Justine A. Smith et al., "Fear of the Human 'Super Predator' Reduces Feeding Time in Large Carnivores," *Proceedings of the Royal Society B* 284 (June 2017): 1–5.

25. Ethan Shaw, "Our Voices Put the Fear in Mountain Lion—And That Puts Them Off Their Food," *Earth Touch News Network*, June 22, 2017, www .earthtouchnews.com/conservation/human-impact/our-voices-put-the -fear-in-mountain-lions-and-that-puts-them-off-their-food/.

26. John W. Laundré and Christopher Papouchis, "The Elephant in the Room: What Can We Learn from California Regarding the Use of Sport Hunting of Pumas (*Puma concolor*) as a Management Tool?," *PlosOne* 15, no. 2 (February 13, 2020): e0224638.

27. Aylin Woodward, "There May Be a Clear Reason a Jogger in Colorado Got Attacked by a Mountain Lion—And It Means Another Attack Could Easily Happen," *Business Insider*, June 1, 2019, www.insider.com/why-mountain -lion-attacked-colorado-jogger-hunting-2019-4.

28. Rico Moore, "Are State Actions Increasing the Risk of Cougars Attacking People?," *Boulder Weekly*, September 12, 2019, www.boulderweekly.com /news/are-state-actions-increasing-the-risk-of-cougars-attacking-people/.

29. Shelby Lin Erdman, "Boy Survives Vicious Cougar Attack; Mom Pries Boy's Arm from Cougar's Jaws," *AJC*, April 3, 2019, https://www.ajc.com

/news/national/boy-survives-vicious-cougar-attack-mom-pries-boy-arm
-from-cougar-jaws/dfF29b6QBmqbd5kXrfd25J/.

30. Mattson et al., "Factors Governing Risk of Cougar Attacks on Humans."

Chapter 3

1. *OED Online* (Oxford, UK: Oxford University Press), www.oed.com,
   accessed December 14, 2008.
2. California Statutes, "Fish and Game Code Section 4800-4809," www
   .suspect.com/laws/California-Codes/Fish-and-Game-Code/sec-4800
   .html, accessed April 2008.
3. Tom Stienstra, "Study Finds Mountain Lions Are Feasting on House Pets,"
   San Francisco Chronicle, February 14, 2016, www.sfgate.com/outdoors
   /article/Study-finds-mountain-lions-are-feasting-on-house-6829205.php.
4. Ben Romans, "Study: Mountain Lions Are Eating Large Number of Pets,"
   Field and Stream, February 25, 2016, www.fieldandstream.com/blogs/field
   -notes/study-mountain-lions-are-eating-large-number-of-pets/.
5. Ryan Sabalow and Phillip Reese, "Why We Still Kill Cougars," *Sacramento Bee*, November 3, 2017, www.sacbee.com/news/california/article
   182397016.html.
6. US Department of Agriculture, Animal and Plant Health Inspection Service, "Death Loss in US Cattle and Calves Due to Predator and Nonpredator
   Causes, 2015," www.aphis.usda.gov/animal_health/nahms/general/down
   loads/cattle_calves_deathloss_2015.pdf, accessed October 1, 2019.
7. Humane Society of the United States, "Government Data Confirm That
   Cougars Have a Negligible Effect on US Cattle & Sheep Industries,"
   www.humanesociety.org/sites/default/files/docs/Cougar-Livestock-6
   .Mar_.19-Final.pdf, accessed October 1, 2019.
8. US Department of Agriculture, Animal and Plant Health Inspection
   Service, "Sheep and Lamb Predator and Nonpredator Death Loss in the
   United States," www.aphis.usda.gov/animal_health/nahms/sheep/down-
   loads/sheepdeath/SheepDeathLoss2015.pdf, accessed October 1, 2019.
9. Matthew W. Alldredge, Frances E. Buderman, and Kevin A. Blecha,
   "Human–Cougar Interactions in the Wildland–Urban Interface of Colorado's Front Range," *Ecology and Evolution* 9, no. 18 (September 2019):
   1–17.
10. Kevin A. Blecha, Randall B. Boone, and Matthew W. Alldredge, "Hunger

Mediates Apex Predator's Risk Avoidance," *Journal of Animal Ecology* 87, no. 3 (May 2018): 609–22.

11. Tom Kuglin, "Montana's Policy against Relocating Mountain Lions Comes under Scrutiny after Helena Incident," *Independent Record*, May 31, 2019,

12. https://helenair.com/outdoors/montana-s-policy-against-relocating -mountain-lions-comes-under-scrutiny/article_96bcb815-8f34-5f67 -9285-32c018a6027e.html.

13. California Department of Fish and Wildlife, "Keep Me Wild: Mountain Lion," www.wildlife.ca.gov/Keep-Me-Wild/Lion, accessed October 12, 2019.

14. Toni K. Ruth et al., "Evaluating Cougar Translocation in New Mexico," *Journal of Wildlife Management* 62, no. 4 (October 1998): 1264–75.

15. Robert C. Belden and Bruce W. Hagedorn, "Feasibility of Translocating Panthers into Northern Florida," *Journal of Wildlife Management* 57, no. 2 (April 1993): 388–97.

16. Francisco E. Fontúrbel and Javier A. Simonetti, "Translocations and Human–Carnivore Conflicts: Problem Solving or Problem Creating?," *Wildlife Biology* 17, no. 2 (June 2011): 217–24.

17. Oregon Department of Fish and Wildlife, "Hunting Cougar in Oregon," https://myodfw.com/articles/hunting-cougar-oregon, accessed August 12, 2019.

18. Wain Evans, *The Cougar in New Mexico: Biology, Status, Depredation of Livestock, and Management Recommendations* (Santa Fe, NM: New Mexico Department of Game and Fish, 1993).

19. Kaylie A. Peebles et al., "Effects of Remedial Sport Hunting on Cougar Complaints and Livestock Depredations," *PLoS ONE* 8, no. 11 (November 19, 2013): e79713.

20. "Encounters between Humans and Mountain Lions Are on the Rise, and Experts Tell Us Why," *ABC News*, April 6, 2019, www.610kona.com /encounters-between-humans-and-mountain-lions-are-on-the-rise-and -experts-tell-us-why/.

21. Kristine J. Teichman, Bogdan Cristescu, and Chris T. Darimont, "Hunting as a Management Tool? Cougar–Human Conflict Is Positively Related to Trophy Hunting," *BMC Ecology* 16 (October 11, 2016): 44.

22. Omar Ohrens, Cristián Bonacic, and Adrian Treves, "Non-Lethal Defense

of Livestock against Predators: Flashing Lights Deter Puma Attacks in Chile," *Frontiers in Ecology and the Environment* 17, no. 1 (February 2019): 32–38.

Chapter 4

1. Maurice G. Hornocker, "An Analysis of Mountain Lion Predation upon Mule Deer and Elk in the Idaho Primitive Area," *Wildlife Monographs* 21 (March 1970): 1–39.
2. David J. Mattson, "State-Level Management of a Common Charismatic Predator: Mountain Lions of the West," in *Large Carnivore Conservation: Integrating Science and Policy in North America*, eds. Susan G. Clark and Murray B. Rutherford (Chicago: University of Chicago Press, 2014), 29–64.
3. L. Mark Elbroch, Jennifer Feltner, and Howard Quigley, "Human–Carnivore Competition for Antlered Ungulates: Do Pumas Select for Bulls and Bucks?," *Wildlife Research* 44, nos. 6–7 (December 2017): 523–33.
4. L. Mark Elbroch, Jennifer Feltner, and Howard Quigley, "Stage-Dependent Puma Predation on Dangerous Prey," *Journal of Zoology* 302, no. 3 (July 2017): 164–70.
5. Blake Lowrey, L. Mark Elbroch, and Len Broberg, "Is Individual Prey Selection Driven by Chance or Choice? A Case Study in Cougars (*Puma concolor*)," *Mammal Research* 61 (August 31, 2016): 353–59.
6. L. Mark Elbroch et al., "The Difference between Killing and Eating: Ecological Shortcomings of Puma Energetic Models," *Ecosphere* 5, no. 5 (May 15, 2014): 1–16.
7. Jedediah Brodie et al., "Relative Influence of Human Harvest, Carnivores, and Weather on Adult Female Elk Survival across Western North America," *Journal of Applied Ecology* 50, no. 2 (April 2013): 295–305.
8. Paul M. Lukacs et al., "Factors Influencing Elk Recruitment across Ecotypes in the Western United States," *Journal of Wildlife Management* 82, no. 4 (May 2018): 698–710.
9. Tavis Forrester and Heiko Wittmer, "A Review of the Population Dynamics of Mule Deer and Black-Tailed Deer *Odocoileus hemionus* in North America," *Mammal Review* 43, no. 4 (October 2013): 292–308.
10. Mark A. Hurley et al., "Demographic Response of Mule Deer to

Experimental Reduction of Coyotes and Mountain Lions in Southeastern Idaho," *Wildlife Monographs* 178, no. 1 (August 2011): 1–33.

11. Eric M. Rominger and Elise J. Goldstein, *Evaluation of an 8-Year Mountain Lion Removal Management Action on Endangered Desert Bighorn Sheep Recovery* (Santa Fe, NM: New Mexico Department of Game and Fish, 2008).

12. Eric M. Rominger, "Culling Mountain Lions to Protect Ungulate Populations—Some Lives Are More Sacred than Others," in *Transactions of the 72nd North American Wildlife and Natural Resources Conference* (Washington, DC: Wildlife Management Institute, 2007), 186–93.

13. Heather E. Johnson et al., "Evaluating Apparent Competition in Limiting the Recovery of an Endangered Ungulate," *Oecologia* 171 (2013): 295–307.

14. Joel Bergerand and John D. Wehausen, "Consequences of a Mammalian Predator–Prey Disequilibrium in the Great Basin," *Conservation Biology* 5, no. 2 (June 1991): 244–48.

15. Julia Rosen, "The Cost of the Bighorn Comeback," *High Country News*, May 29, 2017, www.hcn.org/issues/49.9/Wildlife-Services-mountain-lion-killing.

16. Charles Creekmore, "The Paradox Called Cougar," *National Wildlife Magazine*, December 1, 1991, www.nwf.org/en/Magazines/National-Wildlife/1992/The-Paradox-Called-Cougar.

17. New Mexico Department of Game and Fish, "Bighorn Sheep Biologist Awarded Prestigious Honor," www.wildlife.state.nm.us/bighorn-sheep-biologist-awarded-prestigious-honor/, accessed on August 12, 2019.

18. Marco Festa-Bianchet et al., "Stochastic Predation Events and Population Persistence in Bighorn Sheep," *Proceedings of the Royal Society B: Biological Sciences* 273, no. 1593 (March 2006): 1537–43.

19. Mattson, "State-Level Management of a Common Charismatic Predator"; Tucker Murphy and David W. MacDonald, "Pumas and People: Lessons in the Landscape of Tolerance from a Widely Distributed Felid," in *Biology and Conservation of Wild Felids*, eds. David W. MacDonald and Andrew J. Loveridge (Oxford, UK: Oxford University Press, 2010), 431–52.

Chapter 5

1. James W. Cain and Michael S. Mitchell, "Evaluation of Key Scientific Issues in the Report, 'State of the Mountain Lion—A Call to End Trophy Hunting

of America's Lion,'" US Geological Survey Open-File Report 2018-1128, https://pubs.usgs.gov/of/2018/1128/ofr20181128.pdf, accessed May 1, 2019.

2. Michael L. Wolfe et al., "Is Anthropogenic Cougar Mortality Compensated by Changes in Natural Mortality in Utah? Insight from Long-Term Studies," *Biological Conservation* 182 (2015): 187–96.

3. Perry Backus, "Hounded to Death: Veteran Libby Mountain Lion Hunter Wade Nixon Says It's Time for Houndsmen in His Area to Ease Up—Before the Big Cats Fade from Northwest Montana," *The Missoulian*, February 2, 2006, https://missoulian.com/outdoors/hounded-to-death-veteran-libby-mountain-lion-hunter-wade-nixon/article_81c31609-7210-565b-ac50-1e22a17074ce.html.

4. Hillary S. Cooley et al., "Does Hunting Regulate Cougar Populations: A Test of the Compensatory Mortality Hypothesis," Ecology 90, no. 10 (October 2009): 2913–21.

5. Benjamin T. Maletzke et al., "Effects of Hunting on Cougar Spatial Organization," *Ecology and Evolution* 4, no. 11 (June 2014): 2178–85.

6. Chris T. Darimont, Brian F. Codding, and Kristen Hawkes, "Why Men Trophy Hunt," *Biology Letters* 13, no. 3 (March 2017): 20160909.

7. Martin Leclerc et al., "Hunting Promotes Spatial Reorganization and Sexually Selected Infanticide," *Science Reports* 7 (2017): 45222.

8. Craig Packer et al., "Sport Hunting, Predator Control and Conservation of Large Carnivores," *PLoS ONE* 4, no. 6 (June 17, 2009): e5941.

9. Maletzke et al., "Effects of Hunting on Cougar Spatial Organization."

10. L. Mark Elbroch et al., "Adaptive Social Behaviors in a Solitary Carnivore," *Science Advances* 3, no. 10 (October 11, 2017): e1701218.

11. Kenneth Logan and Linda Sweanor, *Desert Puma* (Washington DC: Island Press, 2000).

12. Kenneth Logan, "Structuring Felid Management for Science," *Wild Felid Monitor* 5 (Summer 2012): 11–14.

13. L. Mark Elbroch and Anna L. Kusler, "Are Pumas Subordinate Carnivores, and Does It Matter?," *PeerJ* 6 (January 24, 2018): e4293.

14. L. Mark Elbroch et al., "Impacts of Reintroduced Wolves Trump Human Hunting When Assessing Effects on a Subordinate Apex Predator Outside Protected Areas," (in review).

Chapter 6

1. R. Harvey Lemelin, "Doubting Thomases and the Cougar: The Perceptions of Puma Management in Northern Ontario, Canada," *Sociologia Ruralis* 49, no. 1 (January 2009): 56–69.
2. See: http://coyotes-wolves-cougars.blogspot.com/2011/09/john-lutz-of-eastern-puma-research.html.
3. David Dishneau, "Puma Network Says Government Turns Blind Eye to Eastern Cougars," *Arizona Daily Sun*, June 14, 2001, https://azdailysun.com/puma-network-says-government-turns-blind-eye-to-eastern-cougars/article_84cde324-2dbd-53ad-9a39-1a45651986de.html.
4. John Lutz, "Saturday, September 24, 2011," *Coyotes, Wolves, and Cougars . . . Forever!* (blog), http://coyotes-wolves-cougars.blogspot.com/2011/09/john-lutz-of-eastern-puma-research.html, accessed September 6, 2019.
5. John Platt, "Giving Up on the 'Ghost Cat': Eastern Cougar Subspecies Declared Extinct," *Scientific American*, March 9, 2011, accessed September 6, 2019, https://blogs.scientificamerican.com/extinction-countdown/giving-up-on-the-ghost-cat-eastern-cougar-subspecies-declared-extinct/.
6. Mark McCollough, USFWS Maine Field Office, *Eastern Puma, 5-Year Review* (Orono, ME: US Fish and Wildlife Service, 2011), www.fws.gov/northeast/ecougar/pdf/Easterncougar5-yearreview-final-111610.pdf.
7. Jason E. Hawley et al., "Long-Distance Dispersal of a Subadult Male Cougar from South Dakota to Connecticut Documented with DNA Evidence," *Journal of Mammalogy* 97, no. 5 (September 2016): 1435–40.
8. Michelle A. Larue, Clayton K. Nielsen, and Brent S. Pease, "Increases in Midwestern Cougars Despite Harvest in a Source Population," *Journal of Wildlife Management* 83, no. 6 (August 2019): 1306–13.
9. *Sensu* Davis C. Stoner, "Dispersal Behaviour of a Polygynous Carnivore: Do Cougars *Puma concolor* Follow Source-Sink Predictions?," *Wildlife Biology* 19, no. 3 (September 2013): 289–301.
10. Michelle A. LaRue and Clayton K. Nielsen, "Population Viability of Cougars in Midwestern North America," *Ecological Modelling* 321 (February 2016):121–29.
11. Craig Pittman, "Florida Panthers: Where Are They Now?," *Flamingo*, May 27, 2016, www.flamingomag.com/2016/05/27/florida-panthers-back-from-the-brink/.

12. Liza Gross, "Why Not the Best? How Science Failed the Florida Panther," *PLoS Biology* 3, no. 9 (August 2005): e333.

13. Madelon Van de Kerk et al., "Dynamics, Persistence, and Genetic Management of the Endangered Florida Panther Population," *Wildlife Monographs* 203, no. 1 (July 2019): 3–35.

Chapter 7

1. Jeremy T. Bruskotter and Robyn S. Wilson, "Determining Where the Wild Things Will Be: Using Psychological Theory to Find Tolerance for Large Carnivores," *Conservation Letters* 7, no. 3 (May/June 2014): 158–65.

2. Ibid.

3. Sophie L. Gilbert et al., "Socioeconomic Benefits of Large Carnivore Recolonization through Reduced Wildlife–Vehicle Collisions," *Conservation Letters* 10, no. 4 (July/August 2017): 431–39.

4. Thomas H. Kunz et al., "Ecosystem Services Provided by Bats," *Annals of the New York Academy of Sciences* 1223 (March 2011): 1–38.

5. Nicola Gallai et al., "Economic Valuation of the Vulnerability of World Agriculture Confronted with Pollinator Decline," *Ecological Economics* 68, no. 3 (January 2009): 810–21.

6. Caroline E. Krumm et al., "Mountain Lions Prey Selectively on Prion-Infected Mule Deer," *Biology Letters* 6, no. 2 (October 2009): 209–11.

7. Taal Levi et al., "Deer, Predators, and the Emergence of Lyme Disease," *Proceedings of the National Academy of Sciences* 109, no. 27 (July 2012): 10942–47.

8. Jeremy Kahn, "On the Prowl," *Smithsonian Magazine*, November 2007, www.smithsonianmag.com/science-nature/on-the-prowl-173579183/.

9. John Duffield, Chris Neher, and David Patterson, *Wolves and People in Yellowstone: Impacts on the Regional Economy* (Missoula, MT: University of Montana, 2006), https://defenders.org/sites/default/files/publications /wolves_and_people_in_yellowstone.pdf.

10. L. Mark Elbroch et al., "Contrasting Bobcat Values," *Biodiversity and Conservation* 26 (July 2017): 2987–92.

11. Kevin Carville, "Bodhi Expeditions" (website), www.bodhiexpeditions.com, accessed October 1, 2019.

12. Guillaume Chapron and Adrian Treves, "Blood Does Not Buy Goodwill:

Allowing Culling Increases Poaching of a Large Carnivore," *Proceedings of the Royal Society B* 283 (April 2016): 20152939.

13. Stephen Kellert, "American Attitudes toward and Knowledge of Animals: An Update," in *Advances in Animal Welfare Science 1984*, eds. M. W. Fox and Linda Mickley (Dordrecht, Netherlands: Martinus Nijhoff Publishers, 1985), 177–213.

14. David Peterson, *Heartsblood: Hunting, Spirituality, and Wildness in America* (Boulder, CO: David Peterson Books, 2010).

15. Leslie Patten, *Ghostwalker: Tracking a Mountain Lion's Soul through Science and Story* ([no city listed]: Far Cry Publishing, 2018), 108.

16. L. Mark Elbroch et al., "Vertebrate Diversity Benefiting from Carrion Subsidies Provided by Subordinate Carnivores," *Biological Conservation* 215 (August 2017): 123–31.

17. Ibid.

18. Gus Lubin and Mamta Badkar, "15 Facts about McDonald's That Will Blow Your Mind," *Business Insider*, November 25, 2011, www.businessinsider.com/facts-about-mcdonalds-blow-your-mind-2011-11/#e-only-place-in-the-lower-48-that-is-more-than-100-miles-from-a-mcdonalds-is-a-barren-plain-in-south-dakota-12.

19. Joshua M. Barry et al., "Pumas as Ecosystem Engineers: Ungulate Carcasses Support Beetle Assemblages in the Greater Yellowstone Ecosystem," *Oecologia* 189 (2019): 577–86.

20. Michel Kohl et al., "Do Prey Select for Vacant Hunting Domains to Minimize a Multi-Predator Threat?," *Ecology Letters* 22, no. 11 (November 2019), 1724–33.

21. John Laundré, *Phantoms of the Prairie: The Return of Cougars to the Midwest* (Madison, WI: University of Wisconsin Press, 2012), 31.

22. Susan K. Jacobson et al., "Content Analysis of Newspaper Coverage of the Florida Panther," *Conservation Biology* 26, no. 1 (February 2012): 171–79.

23. Saloni Bhatia et al., "Understanding the Role of Representations of Human–Leopard Conflict in Mumbai through Media-Content Analysis," *Conservation Biology* 27, no. 3 (June 2013): 588–94.

24. R. Harvey Lemelin, "Doubting Thomases and the Cougar: The Perceptions of Puma Management in Northern Ontario, Canada," *Sociologia Ruralis* 49, no. 1 (January 2009): 56–69.

25. David J. Mattson, "State-Level Management of a Common Charismatic Predator: Mountain Lions of the West," in *Large Carnivore Conservation: Integrating Science and Policy in North America*, eds. Susan G. Clark and Murray B. Rutherford (Chicago: University of Chicago Press, 2014), 29–64.

Chapter 8

1. "Region 3 Mountain Lion Season Meeting, Three Forks, Montana, March 13, 2014," archived by Enhancing Montana's Wildlife and Habitat, accessed August 15, 2019, www.emwh.org/issues/predators/mar%20 2014%20%20proffitt%20mt%20lion%20study.mp3.
2. Kelly M. Proffitt et al., "Integrating Resource Selection into Spatial Capture-Recapture Models for Large Carnivores," *Ecosphere* 6 (November 20, 2015): 239.
3. Hugh S. Robinson and Richard M. DeSimone, *The Garnet Range Mountain Lion Study: Characteristics of a Hunted Population in West-central Montana, Final Report* (Helena, MT: Montana Department of Fish, Wildlife and Parks, 2011).
4. Daniel R. Eaker et al., "Annual Elk Calf Survival in a Multiple Carnivore System," *Journal of Wildlife Management* 80, no. 8 (2016): 1345–59.
5. Brett French, "Hunters Criticize Mountain Lion Quotas in Bitterroot Districts," *Billings Gazette*, April 14, 2014, https://ravallirepublic.com/news /local/article_ab1c9d33-65d7-5d60-b309-d1325f4059e0.html.
6. Michael S. Mitchell et al., "Distinguishing Values from Science in Decision Making: Setting Harvest Quotas for Mountain Lions in Montana," *Journal of Wildlife Management* 42, no. 1 (March 2018): 13–21.
7. John F. Organ and Richard E. McCabe, "History of State Wildlife Management in the United States," in *State Wildlife Management and Conservation*, ed. Thomas J. Ryder (Baltimore, MD: Johns Hopkins University Press, 2018), 1–23.
8. Kathleen A. Griffin et al., "Neonatal Mortality of Elk Driven by Climate, Predator Phenology, and Predator Community Composition," *Journal of Animal Ecology* 80, no. 6 (November 2011): 1246–57.
9. Eaker et al., "Annual Elk Calf Survival in a Multiple Carnivore System."
10. Michael Thompson, *Outcomes of the Region 2, Montana, Mountain Lion Working Group, May 2014* (Missoula, MT: Montana Department of Fish, Wildlife & Parks, 2014).

11. Ibid.

12. Daniel J. Decker et al., "Moving the Paradigm from Stakeholders to Beneficiaries in Wildlife Management," *Journal of Wildlife Management* 83, no. 3 (April 2019): 513–18.

13. US Department of the Interior, US Fish and Wildlife Service, and US Department of Commerce, US Census Bureau, *2016 National Survey of Fishing, Hunting, and Wildlife-Associated Recreation*, https://wsfrprograms. fws.gov/subpages/nationalsurvey/nat_survey2016.pdf, accessed August 20, 2019.

14. *Giving USA, "The Annual Report on Philanthropy for the Year 2018,"* https://givingusa.org/, accessed October 15, 2019.

15. Nathan Rott, "Decline in Hunters Threatens How U.S. Pays for Conservation," *National Public Radio*, March 20, 2018, www.npr.org/2018/03/20/593001800/decline-in-hunters-threatens-how-u-s-pays-for-conservation.

16. US Fish and Wildlife Service et al., *2016 National Survey of Fishing, Hunting, and Wildlife-Associated Recreation*.

17. Andrew Loftus Consulting & Southwick Associates, "Financial Returns to Industry from the Federal Aid in Wildlife Restoration Program," 2011, https://wsfrprograms.fws.gov/Subpages/GrantPrograms/MultiState/MS_WRTaxReport2011.pdf, accessed August 20, 2019.

18. Willie Robertson and William Doyle, *American Hunter: How Legendary Hunters Shaped America* (Brentwood, TN: Howard Books, 2016).

19. Organ and McCabe, "History of State Wildlife Management in the United States."

20. Adrian Treves, Kyle A. Artelle, and Paul C. Paquet, "Differentiating between Regulation and Hunting as Conservation Interventions," *Conservation Biology* 33, no. 2 (April 2019): 472–75.

21. Kim Parker, "Among Gun Owners, NRA Members Have a Unique Set of Views and Experiences," *FactTank*, July 5, 2017, www.pewresearch.org/fact-tank/2017/07/05/among-gun-owners-nra-members-have-a-unique-set-of-views-and-experiences/.

22. Cynthia A. Jacobson et al., "A Conservation Institution for the 21st Century: Implications for State Wildlife Agencies," *Journal of Wildlife Management* 74, no. 2 (February 2010): 203–9.

23. US House of Representatives, Committee of Resources, *Oversight Hearing before the Subcommittee on Fisheries, Wildlife, Oceans of the Committee of Resources, House of Representatives, One Hundred Fourth Congress, Second Session, on the Fish and Wildlife Service Diversity Funding Initiative Known as "Teaming with Wildlife," June 6, 1996,* Serial No. 104-71 (Washington, DC: US Government Printing Office, 1996).

24. Ibid.

25. Organ and McCabe, "History of State Wildlife Management in the United States."

26. Valerius Geist, "Beware of 'Natural' Wildlife Management," *The Outdoorsman* 5 (July 2004): 4–5.

27. Thomas L. Serfass, Robert P. Brooks, and Jeremy T. Bruskotter, "North American Model of Wildlife Conservation: Empowerment and Exclusivity Hinder Advances in Wildlife Conservation," *Canadian Wildlife and Management* 7, no. 2 (2018): 101–18.

28. John F. Organ et al., *The North American Model of Wildlife Conservation,* The Wildlife Society Technical Review 12-04 (Bethesda, MD: The Wildlife Society, 2012).

29. M. Nils Peterson and Michael Paul Nelson, "Why the North American Model of Wildlife Conservation Is Problematic for Modern Wildlife Management," *Human Dimensions of Wildlife* 22 (September 2016): 43–54.

30. Pope & Young Club, "Rules of Fair Chase," https://pope-young.org/fairchase/default.asp, accessed July 18, 2019.

31. Boone and Crockett Club, "The Principles of Fair Chase," https://www.boone-crockett.org/huntingEthics/ethics_fairchase_principles.asp?area=huntingEthics, accessed July 18, 2019.

32. Organ et al., *The North American Model of Wildlife Conservation.*

33. Wes Siler, "It's Time for Hunters to Leave the NRA," *Outside Magazine,* July 23, 2018, www.outsideonline.com/2328866/its-time-hunters-leave-nra.

34. Daniel J. Herman, *Hunting and the American Imagination* (Washington, DC: Smithsonian Institution Press, 2001).

35. Siler, "It's Time for Hunters to Leave the NRA"

36. National Rifle Association, "The Fight to Save Hunting," https://www.nrahunting.com/, accessed August 20, 2019.

37. Decker et al., "Moving the Paradigm."

38. Gary Yourofsky, "Empathy, Education, and Violence: A Time for Everything," www.adaptt.org/animal-rights/empathy-education-and-violence-a-time-for-everything.html, accessed September 12, 2019.

39. David J. Mattson, "State-Level Management of a Common Charismatic Predator: Mountain Lions of the West," in *Large Carnivore Conservation: Integrating Science and Policy in North America*, eds. Susan G. Clark and Murray B. Rutherford (Chicago: University of Chicago Press, 2014), 29–64.

40. Daniel J. Decker et al., "Collaboration for Community-Based Wildlife Management," *Urban Ecosystems* 8 (June 2005): 227–36.

Chapter 9

1. Matthew J. Warren et al., "Forest Cover Mediates Genetic Connectivity of Northwestern Cougars," *Conservation Genetics* 17 (April 2016): 1011–24.

2. John F. Benson et al., "Extinction Vortex Dynamics of Top Predators Isolated by Urbanization," *Ecological Applications* 29, no. 3 (April 2019): e01868.

3. John W. Laundré and Christopher Papouchis, "The Elephant in the Room: What Can We Learn from California Regarding the Use of Sport Hunting of Pumas (*Puma concolor*) as a Management Tool?," *PlosOne* 15, no. 2 (February 13, 2020): e0224638.

4. Richard A. Beausoleil et al., "Research to Regulation: Cougar Social Behavior as a Guide to Management," *Wildlife Society Bulletin* 37, no. 3 (May 2013): 680–88.

5. Connor O'Malley et al., "Aligning Mountain Lion Hunting Seasons to Mitigate Orphaning Dependent Kittens," *Wildlife Society Bulletin* 42, no. 3 (September 2018): 438–43.

6. Richard A. Beausoleil and Kenneth I. Warheit, "Using DNA to Evaluate Field Identification of Cougar Sex by Agency Staff and Hunters Using Trained Dogs," *Wildlife Society Bulletin* 39, no. 1 (March 2015): 203–9.

7. Elizabeth Byrd, John G. Lee, and Nicole J. Olynk Widmar, "Perceptions of Hunting and Hunters by U.S. Respondents," *Animals* 7, no. 11 (November 2017): 83.

8. Klickitat County Sheriff's Office, "27 August 2019," Facebook posting, www.facebook.com/permalink.php?story_fbid=3036118436429581&id=492259454148838. accessed November 1, 2019.

9. Don Jenkins, "Washington Sheriff Takes the Lead in Pursuing Cougars," *Capital Press*, September 9, 2019, www.capitalpress.com/ag_sectors/livestock /washington-sheriff-takes-lead-in-pursuing-cougars/article_cae2a1a4 -d0e1-11e9-9965-937aa24afc1e.html.

10. Simon Gutierrez, "Rise in Cougars in Klickitat Co. Prompts Posse of Hunters Deputized by the County Sheriff," *Fox 12 Oregon*, November 13, 2019, https://www.kptv.com/news/rise-in-cougars-in-klickitat-co-prompts -posse-of-hunters/article_97836c04-0691-11ea-8595-cf2a8157ca40.html.

11. Cynthia A. Jacobson and Daniel J. Decker, "Governance of State Wildlife Management: Reform and Revive or Resist and Retrench?," *Society and Natural Resources* 21, no. 5 (May 2008): 441–48.

12. John F. Organ and Richard E. McCabe, "History of State Wildlife Management in the United States," in *State Wildlife Management and Conservation*, ed. Thomas J. Ryder (Baltimore, MD: Johns Hopkins University Press, 2018), 1–23.

13. Jacobson and Decker, "Governance of State Wildlife Management."

14. Daniel J. Decker et al., "Moving the Paradigm from Stakeholders to Beneficiaries in Wildlife Management," *Journal of Wildlife Management* 83, no. 3 (April 2019): 513–18.

15. Caleb M. Bryce, Christopher. C. Wilmers, and Terrie M. Williams, "Energetics and Evasion Dynamics of Large Predators and Prey: Pumas vs. Hounds," *PeerJ* 5 (August 2017): e3701.

16. Niccolò Machiavelli, *The Prince* (New York: Penguin Books, 1961), 109.

About the Author

MARK ELBROCH is a mountain lion biologist and director of Panthera's Puma (Mountain Lion) Program. He has contributed to mountain lion research in Idaho, California, Colorado, Wyoming, Washington, Mexico, and Chile. Mark's work with mountain lions has been covered by *National Geographic*, the BBC, National Public Radio, the *New York Times*, *Scientific American*, and the *Washington Post*, among others. Mark is twice a National Geographic Explorer, a 2011 Robert and Patricia Switzer Foundation Fellow (for environmental leadership), and the 2017 recipient of the Craighead Conservation Award for "creating positive and lasting conservation outcomes." As a working biologist, Mark often finds himself caught in the middle of opposing opinions about mountain lions. On the one hand, he has been called a "soulless bastard" for "enslaving" mountain lions for research, and on the other, a "liberal, tree-hugging, s**t-for-brains" for advocating for reducing mountain lion hunting. This unique combination provides him the distinctive vantage from which he wrote this book. Mark currently lives in lion country with his family, and continues to design and implement mountain lion research and conservation strategies in both North and South America for Panthera (www.panthera.org). He is the author or coauthor of ten guidebooks on natural history. More about Mark can be found at markelbroch.com.

Index